# PARTICLE SIZE ANALYSIS

# PARTICLE SIZE ANALYSIS

edited by

## JOHN D. STOCKHAM

Science Advisor
Chemistry & Chemical Engineering Research Division
and
Manager of Fine Particles Research
IIT Research Institute
Chicago, Illinois

and

## EDWARD G. FOCHTMAN

Engineering Advisor
Chemistry & Chemical Engineering Research Division
Manager of Chemical Engineering
IIT Research Institute
Chicago, Illinois

ANN ARBOR SCIENCE
PUBLISHERS INC
P.O. BOX 1425 • ANN ARBOR, MICH. 48106

Copyright © 1977 by Ann Arbor Science Publishers, Inc.
230 Collingwood, P. O. Box 1425, Ann Arbor, Michigan 48106

Library of Congress Catalog Card No. 77-78323
ISBN   0-250-40189-4

# PREFACE

This volume, dealing with the subject of particle size analysis, was prepared by members of the Fine Particles Research Section, Chemistry and Chemical Engineering Division, IIT Research Institute, Chicago, Illinois. The papers were originally presented at Filtration Day—1975 sponsored by the Midwest Chapter of the Filtration Society of England. The support of the Midwest Chapter is acknowledged with special thanks due to T. Clifford Powell, William Gont, William Kowalski, Arvid Malitor and Allan Gaynor. Mr. Edward Fochtman of IITRI served as general program chairman and John Stockham chaired the technical program.

# TABLE OF CONTENTS

# LIST OF FIGURES

# LIST OF TABLES

# WHAT IS PARTICLE SIZE:
## THE RELATIONSHIP AMONG STATISTICAL DIAMETERS

**John D. Stockham**
   IIT Research Institute
   10 East 35th Street
   Chicago, Illinois 60616

## INTRODUCTION

The size of a particle is that dimension which best characterizes its state of subdivision.  For a spherical, homogeneous
particle, the diameter is that dimension and can be unambigu-
ously used as its size.  Particles of irregular shape can be characterized in various ways.  Various equivalent diameters can be
extracted from techniques that measure different size-dependent
properties such as volume, surface, resistance to motion in a
gas, or power of light scattering.  Conversion of size data
developed from one set of particle properties to another can
lead to significant errors.  Because of this, the particle sizing
technique should, wherever possible, duplicate the process
under evaluation.

## STATISTICAL CONCEPTS

Most fine-particle systems, whether formed by comminution
of a bulk material or grown by accretion, have particle size

distributions that obey the log-normal distribution function. When particle size is plotted as a function of the number of times each size occurs, a skewed particle size distribution is obtained as illustrated in Figure 1. In particle size applications, this distribution is typically skewed to the right. Particle size

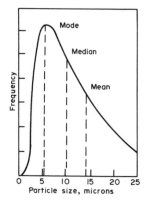

Figure 1. Skewed particle size distribution of typical dust.

distributions are characterized by the parameters that measure the central tendency of the distributions, and the dispersion about the central tendency. There are three measures of central tendency: the average or arithmetic mean, $d_{av}$; the median, $d_{1/2}$; and the mode, $d_o$. The average or mean of the frequency distribution, $d_{av}$, is the center of gravity or balancing point. The main characteristic of the mean is that it is affected by all values actually observed and thus is greatly influenced by extreme values. The median of a frequency distribution, $d_{1/2}$, is the value that divides the frequency distribution into two equal areas. The central tendency of a skewed frequency distribution is more adequately represented by the median rather than the mean. The mode of a distribution is the value that occurs most frequently. It is a seldom-used value in describing particle size distributions.

Because of its desirable mathetmatical properties, the standard deviation, $\sigma$, is almost exclusively used as the measure of dispersion. The standard deviation is the root-mean-square deviation about the mean value. Its derivation and application in significant testing and setting confidence levels can be found in most textbooks on statistics.

When the distribution shown in Figure 1 is replotted using the logarithm of the particle size, the asymmetrical or skewed curve is transformed into a symmetrical or bell-shaped curve, Figure 2. This transformation is of utmost importance because in this form the distribution is amenable to all the statistical

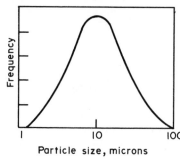

Figure 2. Particle size distribution of Figure 1. Plotted with logarithm of particle size.

procedures developed for the normal or Gaussian distribution. In the log-normal distribution the mean, median and mode coincide and have identical values. The single central tendency value is best termed the geometric median particle size, $d_g$, and the measure of dispersion the geometric standard deviation, $\sigma_g$. These two values completely define the log-normal particle size distribution. Two samples having identical $d_g$ and $\sigma_g$ values can be said to be drawn from the same total population and exhibit properties characteristic of the total population. These two values should be determined and reported as part of any aerosol or powder study. Compliance with the log-normal distribution function can be verified and the values of $d_g$ and $\sigma_g$ easily obtained by plotting the cumulative frequency data on logarithmic probability graph paper (No. 3128, Codex Book Co., Norwood, Mass.). If the particle size distribution obeys the log-normal distribution function, a straight line will be obtained on the log-probability graph as shown in Figure 3. The value of $d_g$ is equal to the 50% value of the distribution, and $\sigma_g$ is equal to the ratio of the 84.1% value divided by the 50% value.

The utility and importance of the log-normal distribution of particle sizes as succinctly summarized by White[1] are:

1. The distribution is completely specified by the two parameters, the geometric median particle size, $d_g$, and the geometric standard deviation, $\sigma_g$.

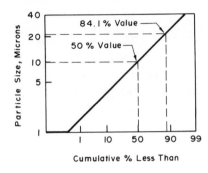

**Figure 3.** Cumulative log-probability curve for particle distribution of Figure 2.

2. The geometric standard deviation is identical for all methods for specifying the particle size distribution, whether by particle number, surface, mass or any other quantity of the form $\kappa d^n$, where d is the diameter and $\kappa$ is a parameter common to all particles. Plots of the cumulative distribution on log-probability paper are then parallel straight lines for number, mass, or surface which leads to a great simplification and simple graphical technique.

3. Transformations among the various particle size parameters and statistical diameters are greatly facilitated both analytically and graphically.

4. Values for the arithmetic mean, median and mode are easily calculated.

$$d_{av} = \text{mean diameter} = \text{antilog}$$
$$(\log d_g + 1.151 \log^2 \sigma_g)$$

$$d_{1/2} = \text{median diameter} = d_g$$

$$d_o = \text{mode diameter} = \text{antilog}$$
$$(\log d_g - 2.303 \log^2 \sigma_g)$$

5. The geometric mean diameter, $d_g$, and the geometric standard deviation, $\sigma_g$, may be found by a simple graphical procedure as illustrated in Figure 3.

6. The geometric mean diameter, $d_g$, is equal to the median or central value of the distribution.

## AVERAGE DIAMETERS

Log-normal particle size distributions can be described completely by two values, $d_g$ and $\sigma_g$, the geometric median diameter and the geometric standard deviation. These values, while completely describing the distribution, have no physical significance to aid in the interpretation of experimental results. Therefore, other average values have been derived to define certain physical properties.

An average diameter is the diameter of a hypothetical particle which in some way represents the total number of particles in the sample. Diameters representing length, surface area, volume, specific surface and falling speed may be determined. Some of the most used average diameters are defined in Table I.

The average diameter that best characterizes the process variable under study should be chosen. For example, the projected area is important for pigments while total surface is important for chemical reactants.[2] Deposition in the respiratory tract is related to $d_w$; the physiological response from materials absorbed on the particles may be related to $d_{vs}$. Thus, aerosols used for medical purposes may have to be described by several parameters.

Table I. Mathematical Definition of Average Diameters

| Average Diameter | Symbol | Mathematical Definition | Description |
|---|---|---|---|
| Arithmetic Mean | $d_{av}$ | $\Sigma nd/\Sigma n$ | The sum of all diameters divided by the total number of particles |
| Surface Mean | $d_s$ | $\sqrt{\Sigma nd^2/\Sigma n}$ | The diameter of a hypothetical particle having average surface area |
| Volume Mean | $d_v$ | $\sqrt[3]{\Sigma nd^3/\Sigma n}$ | The diameter of a hypothetical particle having average volume. The median value of this frequency distribution is often called the mass median diameter. |
| Volume-Surface Mean (also Sauter mean) | $d_{vs}$ | $\Sigma nd^3/\Sigma nd^2$ | The average size based on the specific surface per unit volume |
| Weight Mean Diameter (also DeBroucker mean) | $d_w$ | $\Sigma nd^4/\Sigma nd^3$ | The average size based on the unit weight of the particles |

The calculation of the various diameters from a contrived set of data is illustrated in Table II. Imagine that the data were obtained by counting the number of spherical particles in 13 size categories using a microscope and an eyepiece scalar to measure the diameters. In Figure 4, the distribution given in Table II is plotted on log-probability paper. The

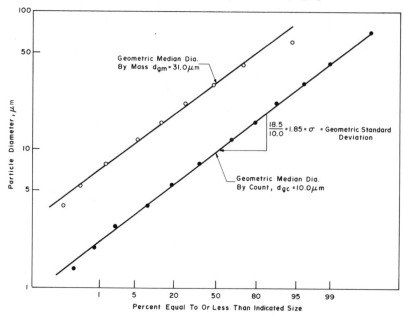

**Figure 4.** Log-probability plot of distribution given in Table 2.

straight lines indicate that the distribution obeys the log-probability distribution function. The use of size categories related by a geometric progression, in this case the $\sqrt{2}$, is advantageous for plotting a log-normal distribution. The geometric median diameter by count or number, $d_{gc}$, is the 50% value of the distribution, 10.0 $\mu$m, and the geometric standard deviation, $\sigma_g$, is the 84.1% value/50% value or 1.85. The additional subscript to the diameter is needed to show that the geometric median diameter measured is based on a number or count frequency. This differentiates it from the geometric median obtained on a mass or surface basis. An additional subscript is not needed for $\sigma_g$ since this value is the

Table II. Calculation of Various Average Diameters

| Particle Size Interval (μm) | Mid Size (μm) d | Freq. of Occurrence n | Cumulative Frequency of n (%) | nd | n log d | nd² | nd³ | Cumulative Frequency of nd³ (%) | nd⁴ |
|---|---|---|---|---|---|---|---|---|---|
| 1.0-1.4 | 1.2 | 2 | 0.2 | 2 | 0.16 | 3 | 3 | — | 4 |
| 1.4-2.0 | 1.7 | 5 | 0.7 | 8 | 1.15 | 14 | 25 | — | 42 |
| 2.0-2.8 | 2.4 | 14 | 2.1 | 34 | 5.32 | 81 | 194 | — | 464 |
| 2.8-4.0 | 3.2 | 60 | 8.1 | 192 | 30.31 | 614 | 1966 | 0.1 | 6291 |
| 4.0-5.6 | 4.8 | 100 | 18.1 | 480 | 68.12 | 2304 | 11059 | 0.3 | 53084 |
| 5.6-8.0 | 6.8 | 190 | 37.1 | 1292 | 158.18 | 8785 | 59742 | 1.3 | 406246 |
| 8.0-12.0 | 10.0 | 250 | 62.1 | 2500 | 250.00 | 25000 | 250000 | 5.6 | 2500000 |
| 12.0-16.0 | 14.0 | 160 | 78.1 | 2240 | 183.38 | 31360 | 429040 | 13.2 | 6146560 |
| 16.0-22.0 | 19.0 | 110 | 89.1 | 2090 | 140.66 | 39710 | 754490 | 26.3 | 14335310 |
| 22.0-30.0 | 26.0 | 70 | 96.1 | 1820 | 99.04 | 47320 | 1230320 | 47.7 | 31988320 |
| 30.0-42.0 | 36.0 | 28 | 98.9 | 1008 | 43.58 | 36288 | 1306368 | 70.4 | 47029248 |
| 42.0-60.0 | 51.0 | 10 | 99.9 | 510 | 17.08 | 26010 | 1326510 | 93.5 | 67652010 |
| 60.0-84.0 | 72.0 | 1 | 100.0 | 72 | 1.86 | 5184 | 373248 | 100.0 | 26872856 |
| TOTALS | | 1000 | | 12248 | 998.84 | 222673 | 5752965 | | 196991436 |

$$d_{av} = \frac{\Sigma nd}{\Sigma n} = 12.2 \ \mu m$$

$$d_{gc} = \text{anti log} \ \frac{\Sigma n \log d}{\Sigma n} = 10.0 \ \mu m$$

$$d_s = \sqrt{\frac{\Sigma nd^2}{\Sigma n}} = 14.9 \ \mu m$$

$$d_v = \sqrt[3]{\frac{\Sigma nd^3}{\Sigma n}} = 17.9 \ \mu m$$

$$d_{vs} = \frac{\Sigma nd^3}{\Sigma nd^2} = 25.4 \ \mu m$$

$$d_w = \frac{\Sigma nd^4}{\Sigma nd^3} = 34.2 \ \mu m$$

same for all manners of expressing the distribution. Also plotted in Figure 4 is the cumulative frequency of $nd^3$. The geometric median diameter of this distribution is $d_{gm}$, the mass median diameter, and is equal to 31.0 $\mu$m.

The values of $d_{av}$, $d_g$, $d_v$, $d_{vs}$, $d_w$ and $d_{gm}$ can be calculated from $d_{gc}$ and $\sigma_g$ using the Hatch-Choate equations.[3] These transformation equations are given in Table III, and the good agreement between the mean diameters calculated directly from the data and by the transformation equation is shown in Table IV. A monograph[4] to convert $d_{gc}$ to $d_{gm}$ is given in Figure 5. The extreme dependence of the transformation on $\sigma_g$ is illustrated. The ratio of $d_{gm}/d_{gc}$ increases from 5.38 to 835 as $\sigma_g$ increases from 2 to 4. For distributions that deviate from the log-normal distribution function, the transformation equations are invalid and should not be used.

#### Table III. Hatch-Choate Transformation Equations

| To Convert From | To | Use the Relationship |
|---|---|---|
| $d_{gc}$, the geometric median diameter by count (count median diameter)[a] | $d_{gm}$ | $\log d_{gm} = \log d_{gc} + 6.908 \log^2 \sigma_g$ |
| | $d_{av}$ | $\log d_{av} = \log d_{gc} + 1.151 \log^2 \sigma_g$ |
| | $d_s$ | $\log d_s = \log d_{gc} + 2.303 \log^2 \sigma_g$ |
| | $d_v$ | $\log d_v = \log d_{gc} + 3.454 \log^2 \sigma_g$ |
| | $d_{vs}$ | $\log d_{vs} = \log d_{gc} + 5.757 \log^2 \sigma_g$ |
| | $d_w$ | $\log d_w = \log d_{gc} + 8.023 \log^2 \sigma_g$ |
| $d_{gm}$, the geometric median diameter by weight (mass median diameter)[b] | $d_{gc}$ | $\log d_{gc} = \log d_{gm} - 6.908 \log^2 \sigma_g$ |
| | $d_{av}$ | $\log d_{av} = \log d_{gm} - 5.757 \log^2 \sigma_g$ |
| | $d_s$ | $\log d_s = \log d_{gm} - 4.605 \log^2 \sigma_g$ |
| | $d_v$ | $\log d_v = \log d_{gm} - 3.454 \log^2 \sigma_g$ |
| | $d_{vs}$ | $\log d_{vs} = \log d_{gm} - 1.151 \log^2 \sigma_g$ |
| | $d_w$ | $\log d_w = \log d_{gm} + 1.151 \log^2 \sigma_g$ |

[a]$d_{gc}$ is obtained from microscopic counting and light-scattering single particles counters.

[b]$d_{gm}$ is obtained by sieving.

Table IV.   Correlation Between Average Diameters Calculated from
Data and the Transformation Equation

| Diameter | Calculated Directly from Data, Table II, $\mu$m | Calculated from Transformation Equations, $\mu$m |
|---|---|---|
| $d_{av}$, arithmetic mean | 12.2 | 12.1 |
| $d_{gc}$, geometric median by count | 10.0 | 10.0[a] |
| $d_s$, surface mean | 14.9 | 14.6 |
| $d_v$, volume mean | 17.9 | 14.6 |
| $d_{vs}$, volume-surface mean | 25.4 | 25.8 |
| $d_w$, weight mean | 34.2 | 37.4 |
| $d_{gm}$, geometric median by mass | 31.0[a] | 31.0 |

[a]from Figure 4.

With the advent of computers and their general availability
through time-share programs, the advantage of the Hatch-Choate
equations is largely negated.   The various mean diameters can
be computed quickly and accurately.   The need to group
data to simplify calculations is unnecessary.   Thus, the errors
inherent in the use of grouped data are eliminated.

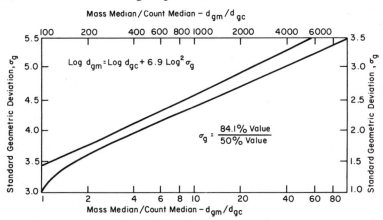

Figure 5.   Graphic estimation of the mass median diameter, $d_{gm}$, from
the count median diameter, $d_{gc}$, and the standard geometric
deviation, $\sigma_g$.

## MEASURING PARTICLE SIZE BY MICROSCOPY

In the foregoing discussion it was tacitly assumed that the particles were spheres the sizes of which were adequately represented by their diameters.

For nonspherical particles the equivalent diameter is the length of any linear intercept that best describes the state of subdivision.  Some of the ways to express the size of nonspherical particles are illustrated in Figure 6.  Feret's diameter

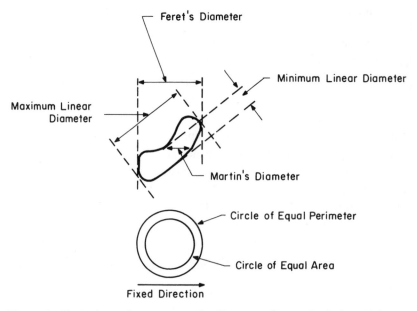

**Figure 6.** Methods used to measure the diameter of nonspherical particles.

is the mean length between two tangents on opposite sides of the particle profile, perpendicular to the fixed direction of scan. Martin's diameter is the mean length of a line, parallel to the fixed direction of scan, that divides the particle profile into two equal areas.  The diameter of a circle of equal area is obtained by estimating the projected area of the particle and comparing it with circles of known diameters scribed in a microscope eyepiece graticule.  Some of the ramifications of the Feret's, Martin's and projected area diameters are discussed by Kaye.[5]  The equal area and Feret's diameter are the most often used.  The area-derived diameter and the Feret's diameter approach one another as the number of particles examined increases. In measuring particle size distributions microscopically, the total number of particles examined should be such that the area and Feret's diameters approach one another.

With many aerosols, interest is often centered on the aerodynamic behavior of the particle. The use of the Stokes' diameter, $d_{st}$, is recommended. The Stokes' diameter is the diameter of a sphere having the same density and free-falling speed as the particle in air. This diameter is obtained from such analytical devices as sedimentometers and cascade impactors which employ aerodynamic principles to classify particles. Size distributions obtained by analyzers that provide number-frequency data must be converted if aerodynamic behavior is sought. Since shape as well as size and density affects aerodynamic behavior, this conversion is tenuous at best. Conversion of $d_{gc}$ to $d_w$ will provide the best approximation to aerodynamic size where specific information on the aerodynamic properties of the particles is lacking.

Caution must be exercised in the use of particle density data. Many fine particles are in reality flocculi and agglomerates having densities well below those reported in literature for the parent material, Table V.

#### Table V. Particle Densities for Agglomerates

| Material | Floc Density g/cc | Normal Density g/cc |
|---|---|---|
| silver | 0.94 | 10.5 |
| mercury | 1.70 | 13.6 |
| cadmium oxide | 0.51 | 6.5 |
| magnesium oxide | 0.35 | 3.65 |
| arsenic trioxide | 0.91 | 3.7 |
| lead monoxide | 0.62 | 9.36 |
| aluminum oxide | 0.18 | 5.57 |
| stannic oxide | 0.25 | 3.70 |
| antimony oxide | 0.63 | 5.57 |
| mercuric chloride | 1.27 | 5.4 |

## CONCLUSIONS

Most fine-particle systems have particle size distributions that follow the log-normal distribution function. These distributions can be completely characterized by two parameters: the geometric median particle size and the geometric standard deviation. These values can be obtained by a simple graphical approach and mathematical transformations are available to convert the geometric median diameter into average diameters that have physical significance.

# 2

# GRAPHICAL PRESENTATION OF
# PARTICLE SIZE DATA

**Earl O. Knutson**
Energy Research & Development Administration
376 Hudson Street
New York, New York 10014

## INTRODUCTION

Data on the size of particles in a particulate system are often presented by means of some type of X-Y plot. When properly prepared, such plots can convey, in a concise manner, much information on particle size.

There are two basic types of X-Y plots for presenting particle size data—the histogram and the cumulative plot— and there are several variants of each. All forms of graphical presentation, however, employ one axis to represent particle size and the other axis to represent particle amount. We shall start with a consideration of these two axes.

## THE PARTICLE SIZE AXIS

The particle size is usually (but not always) represented by the X, or horizontal, axis of the graph paper. The X-axis will be used here.

What, specifically, is plotted on the particle size axis? If the particles in question are spherical, it is certainly logical to plot particle diameter along the size axis. Particle diameters could be determined, for example, from an optical or electron micrograph.

There are many practical situations in which the particles are not spherical. In these cases, the choice of a quantity to represent individual particle size becomes a real challenge. Fortunately, many ways have been devised, each with a certain logic to it. Some of the methods are carried out by non-microscope techniques. The variety of quantities which have been used to represent particle size is shown in Table VI.

**Table VI.   Quantities Used to Represent Size of Individual Particles**

---

Derived from microscopy:
    diameter
    area
    perimeter
    Feret's diameter
    Martin's diameter
    maximum chord
    minimum chord
Derived from other sizing techniques:
    terminal settling velocity (in air)
    Stokes' equivalent diameter
    aerodynamic equivalent diameter
     (unit density sphere diameter)
    equivalent diffusional diameter
    apparent optical diameter
    sieve size

---

Many of the various representations of particle size shown in Table VI will be discussed in other chapters. For the present, it is sufficient to point out that there is a variety of representations, and that the reader should be aware of which representation is being used. It may also be noted that many (but not all) of the measures of particle size may be inter-converted by means of mathematical formulas.

The scale on the particle size axis may be either linear or logarithmic. The logarithmic scale is preferred when the particulate system comprises a broad range of sizes. The linear and logarithmic scales are illustrated in Figure 7.

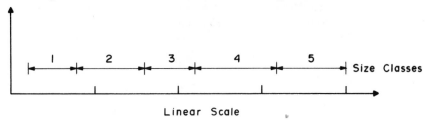

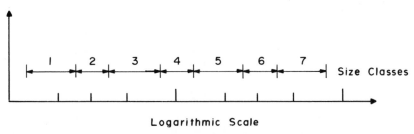

**Figure 7.** Linear and logarithmic particle size scales, with example size classes.

When dealing with large numbers of particles, which is the usual case, it is convenient to subdivide the particle size axis into *size classes.* Example size classes are illustrated in Figure 7. The classes need not be of equal width, but it usually facilitates matters if they are.

In some cases, the size classes to be used on the particle size axis are prescribed by the method used to obtain the particle size information. For example, if a sieve analysis is used, the size class boundaries should be chosen to coincide with the mesh opening of the sieves. These are given in Table VII. It is seen that the mesh sizes progress in the ratio $\sqrt{2}$. These classes would form equal-width classes on the logarithmic particle size scale. Similarly, the Porton graticule, sometimes used to facilitate sizing particles by microscope, has comparison circles whose diameters progress in the ratio $\sqrt{2}$. Several instruments used in sizing particles are designed to obtain equal-width classes on a logarithmic particle size scale.

Table VII. Tyler Sieve Series Mesh Openings

| Tyler Mesh Number | Opening Size ($\mu$m) | |
|---|---|---|
| 400 | 37 | |
| 325 | 44 | |
| 270 | 52 | ratio to |
| 250 | 62 | adjacent |
| 200 | 74 | size = 2¼ |
| 170 | 88 | |
| 150 | 105 | |
| –– | –– | |
| 100 | 148 | |
| –– | –– | |
| 65 | 209 | |
| –– | –– | ratio to |
| 48 | 296 | adjacent |
| –– | –– | size = $\sqrt{2}$ |
| 35 | 419 | |
| –– | –– | |
| 28 | 592 | |
| –– | –– | |
| –– | –– | |
| –– | –– | |

## THE PARTICLE "AMOUNT" AXIS

The particle "amount" axis is used to represent the amount of particulate matter which belongs to specified size classes on the particle size axis. We shall use the Y-axis to represent particle amount.

As with particle size, there are several acceptable methods for representing particle amount. The two most common are: number of particles and mass of particles. If the data being plotted were obtained from a microscope count, it is obvious that number is the most convenient way to specify particle amount. On the other hand, for data obtained by a sieve analysis, the mass is the most convenient way to specify particle amount per size class.

It is worth emphasizing here that the number and the mass of particles per size class are both acceptable methods of specifying particle amount. Both are in common use. The graphs can be very different in appearance. The first question the reader should ask when scrutinizing a particle size distribution plot is: "Are these data in terms of number or in terms of mass?"

Other representations of particle amount include the surface area (important when the particles are involved in surface-controlled chemical reactions) and volume (closely related to mass).

The particle amount axis is usually a linear scale. A logarithmic scale is used on rare occasion. If the particle amount is presented in cumulative form (see next section), a probit scale may be used.

## HISTOGRAMS AND CUMULATIVE PLOTS

The two main forms of graph for presenting particle size data are the *histogram* and the *cumulative plot*. The two differ only in the way the particle "amount" is presented. The particle size axis can be the same for both.

Example histogram and cumulative plots are given in Figure 8. Note that the histogram is a bar graph in which the base of each rectangular bar is the width of the corresponding size class. In a properly drawn histogram, the *area* of each bar is proportional to the amount of particles (in terms of number or mass) within the size class. This is most easily done if the size classes are of equal width, that is, if the class boundaries are in an arithmetic progression for a linear size axis or a geometric progression for a logarithmic size axis.

In the cumulative plot, Figure 8b, points are entered showing the amount of particulate material (in number or mass) contributed by the particles below the specified size. Hence, the curve has a continually rising character.

No specific scales have been indicated for the axes in Figure 8. As mentioned before, the particle size axis (X-axis) can be either linear or logarithmic. The particle amount axis (Y-axis) is usually linear, except as discussed below.

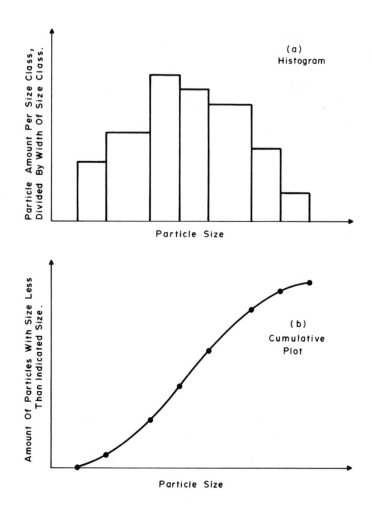

**Figure 8.**  Histogram and cumulative plot.

It is customary to express the Y-axis in the cumulative plot in terms of percent, with 100% representing the total "amount" of particles in the sample.  Thus, the scale on the Y-axis usually runs from 0 to 100%.  Sometimes a linear scale is used, but frequently a probit scale is employed.  The probit scale, illustrated in Figure 9, helps to present particle size data in the form of a straight line.

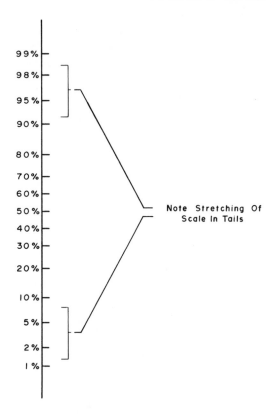

Figure 9.   Probit scale.

## GRAPHICALLY DERIVED
## PARTICLE SIZE CHARACTERISTICS

Even when a particulate system includes a broad range of particle sizes, it is often convenient to speak of a single characteristic size for the system. Many characteristic sizes have been proposed. Most involve a mathematical formula. There is one important characteristic size, however, which can be read off any cumulative plot of the particle size data. This is the *median* particle size.

The median particle size is illustrated in Figure 10. It is defined as that particle size for which the particle amount equals 50% of the total. If the particle amount is represented by the number of particles, then the 50% size is called the

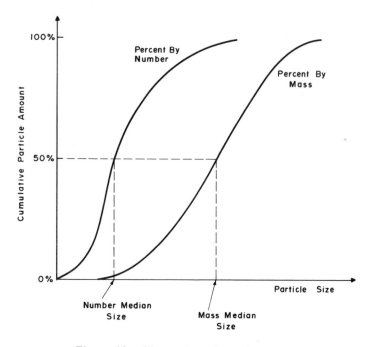

**Figure 10.**   Illustration of median sizes.

*number median* size.   If mass is used as the measure of particle amount, the corresponding median is the *mass median* size. The distinction between number and mass median is very important, since they often differ by a very substantial amount.

As was previously discussed, there are also several acceptable ways of representing the size of individual particles.   The two most common are the actual diameter and the aerodynamic equivalent diameter.   Thus, there are four commonly used median diameters.   These are:

| | |
|---|---|
| the number median diameter (or count median diameter) | NMD (or CMD) |
| the number median aerodynamic equivalent diameter | NMAD |
| the mass median diameter | MMD |
| the mass median aerodynamic equivalent diameter | MMAD |

## SUMMARY

The two main graphical presentations of particle size data are the histogram and the cumulative plot.  In either case, one axis is used to represent particle size, and the other is used to represent particle amount.

The most important question to ask when viewing graphical particle size data is:  "Is the particle amount given in terms of number of particles or mass  of particles?"  This often makes an enormous difference on the meaning of the graph.

The second question to be asked is:  "Is the particle size represented by actual diameter or aerodynamic equivalent diameter?"  This is less important than the previous question, but a misunderstanding on this point can lead to an error of up to a factor of 2.

There are many books or articles which may be consulted for further information on presenting particle size data.  Two good ones[6,7] give numerical examples.

# 3

# SIZING PARTICLES USING THE MICROSCOPE

**George Yamate and John D. Stockham**
IIT Research Institute
10 West 35th Street
Chicago, Illinois 60616

## INTRODUCTION

The microscope is the basic instrument for sizing particles. Any laboratory concerned with particle size measurements should make their initial investment in a microscope. While microscopical sizing has been superseded in recent years by faster and more automated techniques, such as electrical resistance zone and light-scattering particle counters, it remains the referee technique and the only instrument which has the following features and capabilities:

- Ability to examine extremely small sample sizes—on the order of 100 particles.
- Acquire information on size, shape, morphology, color, refractive index, melting point and crystallography of particles.
- Identify particulate matter and often deduce its source.
- Provide a permanent picture of the sample through photomicrography.
- Direct observation of the particles gives a certain degree of credibility and confidence in the results.
- Direct observation permitting the microscopist to obtain or deduce information, such as the presence of agglomerates, hollow bodies

23

or other forms of nonhomogeneity, unavailable from other particle size instruments.

- Low equipment costs—the basic equipment needs for particle sizing can be acquired for less than $2,000—making the microscope a very inexpensive sizing tool.

## MICROSCOPE ESSENTIALS

For particle sizing by microscopy, equipment needs are few and relatively inexpensive. In addition to a basic microscope, the essential components and auxiliary needs are:

- A mechanical X-Y stage with a 3:1 microscope slide holder.
- A minimum of a 10X and 43X achromatic or apochromatic lenses with high numerical apertures.
- Several oculars or eyepieces; most useful will be a 10X, 15X and 20X.
- A stage micrometer.
- Several eyepiece scalars.
- A good source of illumination.

A mechanical stage is essential. One with an X-Y movement is superior to the circular stages common to biological or polarizing microscopes. A micrometer scale on the stage is of great value. It is useful in measuring the area of the deposit scanned, locating preselected randomized positions for counting and in relocating areas of interest.

The resolving power of the microscope is related to the numerical aperture of the objective lens. The resolving power is defined as the ability to distinguish separate details of closely spaced microscopic structures. The theoretical limit of resolving two discrete points a distance X apart is:

$$X = \frac{1.22\lambda}{2NA}$$

where $\lambda$ is the wavelength of length  and NA is the numerical aperture of the objective.

By substituting a wavelength of 4500 Å for visible light and a numeric aperture of 1.3, the best that can be done with visible light, we see that the resolving power of the optical microscope is about 0.2 $\mu$m.

The magnification of the microscope is the product of the objective-eyepiece combination. As a rule of thumb, the maximum useful magnification for the microscope is 1000 times the numerical aperture, Table VIII. Any additional magnification will not improve the resolving power. As shown in Table VIII, the 10X eyepiece, furnished with most microscopes, under-utilizes the capabilities of the most commonly used objective lenses. The 10X eyepiece, therefore, is useful for scanning or crude work, but 15 to 30X eyepieces are needed for detail and critical work.

Table VIII. Maximum Useful Magnification and the Eyepiece Required for Different Objectives

| Objective | | | | Maximum | |
|---|---|---|---|---|---|
| Magnification | Focal Length | Numerical Aperture | Depth of Focus | Useful Magnification | Eyepiece Required |
| 2.5 X | 56 mm | 0.08 | 50 $\mu$m | 80 | 30X |
| 10 X | 16 mm | 0.25 | 8 $\mu$m | 250 | 25X |
| 20 X | 8 mm | 0.50 | 2 $\mu$m | 500 | 25X |
| 43 X | 4 mm | 0.66 | 1 $\mu$m | 660 | 15X |
| 97 X | 2 mm | 1.25 | 0.4 $\mu$m | 1250 | 10X |

The magnification of the objective lens is calculated by dividing the microscope tube length, usually 160 mm, by the focal length.

Depth of field is often a problem to the microscopist attempting to size a powder with a broad spectrum of sizes. Table VIII illustrates how the depth of focus is dependent on the choice of the objective lens. It is recommended that the highest focal length objective possible be used and the magnification increased through the use of a higher power ocular.

The stage micrometer is required to calibrate the eyepiece scaler for making particle size measurements. The microscope must be calibrated for each objective-eyepiece-body tube length combination. Body tubes are often adjustable and peculiar to a manufacturer. Thus, objectives often are not interchangeable among microscopes from different manufacturers.

Particle sizing by microscopy is accomplished with the eyepiece graticules and scalers. Many types are available.[8] The most useful are the linear graduated scale and the Patterson-Cawood and Porton globe-and-circle graticules, Figure 11. The Patterson-Cawood graticule has a series of discs graduated in an arithmetic series; the Porton graticule's discs are graduated

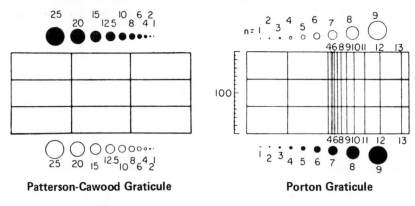

Figure 11. Eyepiece scalars and graticules.

in a series based on the $\sqrt{2}$. The series based on the $\sqrt{2}$ has certain advantages for irregular particles, and it is generally more satisfactory to compare areas than diameters; successive discs double in area as they progress in size. Bosanquet[9] has shown that with the geometric series, the error due to assigning a particle to the wrong class is only 0.18 times the counting error. Moreover, experience has shown that for most distributions, the size range covered by the arithmetic series is too narrow. The geometric series covers a range of 128 to 1.

Illumination is critical—the requirement for a good illumination system is to have uniform intensity of illumination over the entire field of view with independent control of intensity and the angular aperture of the illuminating cone. Köhler illumination is suggested and is described in most texts on microscopy.[10,11]

## MEASUREMENT OF SIZE

Arbitrary techniques are used to size particles by microscopy. For spherical particles, the diameter is the appropriate measurement. For irregular particles, Feret's, Martin's, projected area and maximum horizontal intercept are often employed, Figure 12. One must be consistent in the measurement used and

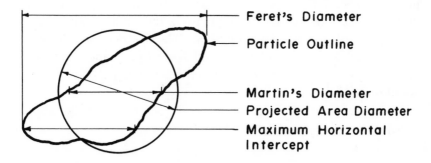

Feret's Diameter

Particle Outline

Martin's Diameter
Projected Area Diameter
Maximum Horizontal Intercept

Figure 12. Various statistical diameters.

always report the method employed. Feret's diameter is the length between two tangents on opposite sides of the particle profile, perpendicular to the fixed direction of scan. It is the easiest to use. Martin's diameter is the length of a line, parallel to the fixed direction of scan, that divides the particle into two equal areas. The projected area diameter is the diameter of a circle having an area equal to the projected area of the particle. The Porton graticule is more useful in estimating the projected area diameter. The maximum horizontal intercept is the longest diameter from particle edge to edge in the direction of scan. It is obtained with image shearing devices and always equals or falls between the Feret's and Martin's diameter.

The determination of a particle size distribution is carried out by measuring the size of a number of particles. Routinely, particles are sized as they are traversed past the eyepiece micrometer by means of the mechanical stage. Each particle, presented in a fixed area of the eyepiece, is sized and tallied into a size class. The eyepiece, objective and size classes are chosen so that particles can be tallied into at least six size classes. The number of particles actually counted may range from 100 to several thousand and depends on the regularity of particle shape, the range of sizes present, the eventual use of the data and the accuracy required. A blood cell counter is a useful device for tallying particles into size classes. After completing the counting and sizing operation, the data are usually plotted on a size-frequency graph for interpretation and the calculation of statistical parameters. These procedures have been described by Stockham[12] and Knutson,[13] and detailed discussion can be found in most texts on fine particle technology.[10,11,14]

## COUNTING REQUIREMENTS

Time and operation fatigue usually limit the number of particles sized by microscopy. However, if certain levels of accuracy are to be achieved, counting and sizing must be continued until the requirements are satisfied. The American Society of Testing and Materials[15] recommends that the number of particles measured in the modal size class should not be less than 25 and should contain at least 100 particles; and at least 10 particles should be present in each size class.

To improve reliability and yet minimize the number of measurements, a statistical technique called "stratified sampling" is employed. Sichel[9] applied the technique to particle sizing and called it "truncated multiple traversing." Sichel's basic criterion is that at least 10 particles must be observed in every size class that has a significant influence on the size curve. It is illustrated in Table IX. In the first traverse, sufficient

Table IX. Data Tabulation for Particle Sizing by Truncated Multiple Traversing System

| Traverse Number | Size Range (μm) | | | | | | | | | | Total Number |
|---|---|---|---|---|---|---|---|---|---|---|---|
| | ⩽0.44 | 0.45-0.62 | 0.63-0.88 | 0.89-1.25 | 1.26-1.76 | 1.77-2.50 | 2.51-3.53 | 3.54-5.00 | 5.01-7.07 | 7.08-10.0 | |
| 1 | 57 | 87 | 54 | 36 | 21 | 24 | 6 | 12 | 0 | 3 | 300 |
| 2 | | | | | | | 12 | 6 | 3 | 0 | 21 |
| 3 | | | | | | | | | 3 | 0 | 3 |
| 4 | | | | | | | | | 2 | 1 | 3 |
| 5 | | | | | | | | | 2 | 1 | 3 |
| 6 | | | | | | | | | 1 | 2 | 3 |
| Total | 57 | 87 | 54 | 36 | 21 | 24 | 18 | 18 | 11 | 7 | 333 |
| Number per traverse | 57 | 87 | 54 | 36 | 21 | 24 | 9 | 9 | 1.8 | 1.2 | 300 |
| Cumulative percent | 19 | 48 | 66 | 78 | 84 | 92 | 95 | 98 | 99.4 | 100 | |

measurements were obtained through size class 2.5 $\mu$m. Therefore, a second traverse, along a new diameter of the particle preparation but of the same area as the first, was made except that only particles greater than 2.5 $\mu$m were counted. The search was continued in a similar manner until Sichel's criteria was satisfied by all size classes. The stratified sampling technique resulted in the sizing of only 333 particles. Without stratified sampling, 1800 particles would have had to be sized to meet the same requirement for accuracy.

Extreme care must be exercised when deriving weight data from size-by-count data as obtained by the microscope. If the transformation equation $\log D_{mmd} = \log D_{nmd} + 6.90 \log^2 \sigma_g$ [12] is to be used, the counting guidelines in Table X must be followed. Therefore, size analyses by the microscope are not favored by the dust collector manufacturers, whose performance standards are given as a weight efficiency. The dust collector industry has

Table X. Number of Particles to be Counted to Achieve any Given Accuracy

| Weight Percentage of Particles In Any Size Range: | 2 | 5 | 10 | 15 | 20 |
|---|---|---|---|---|---|
| Expected Accuracy % | | Number of Particles to be Counted | | | |
| 2 | (3) | (8) | 25 | 56 | 100 |
| 1 | (6) | 25 | 100 | 225 | 400 |
| 0.5 | 16 | 100 | 400 | 900 | 1600 |
| 0.2 | 100 | 625 | 2500 | 5600 | 10000 |
| 0.1 | 400 | 2500 | 1000 | 22500 | 40000 |

agreed to use the results of the Bahco centrifugal air elutriation device [16] for particle size analyses. It is obvious from Table X why image analysis equipment has advanced in the past decade.

## SAMPLE PREPARATION

As with any sizing device, sampling and sample preparation are the keys to success. Because of the countless varieties of samples to be sized, a myriad of sample preparation techniques has been devised. Obviously, a discussion of all the techniques is beyond the scope of this paper. The important considerations of any sampling and sample preparation procedure are that they be unbiased, that the particles represent the total population from which they were obtained and are dispersed randomly without reference to size and shape, that the agglomerates are deflocculated, but individual particles are not shattered by the process of deflocculating, and the areal particle concentration be convenient for sizing. A helpful suggestion is to have 30 to 50 particles per field of view.

As a first approach, the following procedures are suggested for use with bulk powder samples and with samples collected on membrane filters.

### Bulk Powders

A small amount of powder (about a milligram) is taken from the powder with a glass needle or spoon. Steel needles are to be avoided because of possible sample bias with magnetic dusts or powders. The specimen is introduced into a drop of mounting fluid, Table XI, previously placed on a glass slide. The index of refraction of the mounting fluid should differ from that of the particles by at least 0.02; matching of the refractive indexes renders the particles invisible. A cover slip is carefully placed over the mounting fluid and allowed to settle to the surface of the slide. Then, with the eraser end of a pencil gentle pressure is applied to the cover slip using a rotary motion. This rotating motion provides a gentle shearing action that will disperse and deflocculate the particles without crushing or shattering the individual particles. Care must also be exercised so that the fines do not preferentially stream to the edge of the cover slip. Mounting fluids are available in a range of viscosities in addition to refractive indexes. The more viscous fluids retard Brownian motion and prevent rapid reagglomeration. The shearing forces are increased materially by the use of a viscous media; mounting media that set are useful for preparing samples of magnetic and very fine powders.

Table XI. Refractive Indices of Mounting Media Suitable for Dust Examination[9]

| Name | Common Solvent | Refractive Index[a] (25°C) |
|---|---|---|
| Water | — — | 1.333 |
| Silicone oil (30,000 centistokes) | Carbon tetrachloride | 1.404 |
| Glycerin | Water | 1.463 |
| Isobutyl methacrylate | Xylene | 1.47 |
| Mineral oil | Kerosene | 1.48 |
| Cedarwood oil (thick) | Xylene | 1.515 |
| Gum dammar | Turpentine | 1.521 |
| Canada balsam | Xylene | 1.535 |
| Colophony | Turpentine | 1.545 |
| Styrax | Xylene | 1.62+ |
| Aroclor (chlorinated diphenyl) | Xylene | 1.63 |
| Balsam of Tolu | Xylene | 1.64 |
| Polyvinyl alcohol[b] | Water | 1.54 |
| Gelatin | Water | 1.516-1.534 |
| Glucose-pectin | Water | 1.43 |
| Glycerin boriphosphate (Abopon) | Water | 1.44 |
| Polyvinyl acetate[b] | — — | 1.47 |
| Polystyrene[b] | — — | 1.6 |
| Methylene iodide | — — | 1.74 |

[a]Refractive index of resins refers to dried materials. In liquid form, these resins have a lower index, since most solvents have an index considerably lower than that of the resin.

[b]These materials form a hard film that may be used with an oil immersion objective without a coverglass.

## Collected Filter Samples

Samples collected on a filter can be examined under incident light by "clearing" the filter. "Clearing" is accomplished by wetting the back side of the filter with a liquid of the same refractive index as the filter material.

## CONCLUSIONS

The optical microscope remains the basic instrument for particle sizing. It should be the first investment for any laboratory interested in performing size analysis. In addition to particle size, the microscope can provide information on shape, color, refractive index and other important physical, chemical and morphological characteristics that can lead to the identity of the particles and clues as to their source.

# 4

# SIZING WITH MODERN IMAGE ANALYZERS

**Jean Graf**
IIT Research Institute
10 West 35th Street
Chicago, Illinois 60616

## INTRODUCTION

Particle size is an essential parameter to determine when dealing with particle-fluid systems. If particle behavior in a fluid is to be adequately predicted, the factors which influence particle behavior must be known; particle size is one of these factors. Physical size determines how rapidly a particle will settle. The actual physical dimensions—length, width and height of a particle also govern a particle's movement or resistance to movement through a fluid system.

In fluid systems where particles are an undesirable contaminant, it is especially necessary to determine the size of the contaminant particles in order to assess the particle's potential for damage and to effectively control their presence. For example, in machinery cooling and lubricating fluid systems, certain size particles cause frictional wear of moving parts. To be able to control this wear, the contaminant particle size causing the wear must be known in order to chose the proper particulate removal device, usually a filter, which will effectively eliminate the damaging particles with minimal alteration of the normal machinery fluid system operation. It is pointless to install a filter to remove 99% of all particle sizes, reducing the fluid flow rate by over 50%, when the potentially damaging particles are of a size which can be removed by a filter causing only a 10% drop in flow rate. Obviously, the size of the detrimental particles and the efficiency of a filter for removing those particles must be known.

A variety of instruments and techniques is available for determining particle size distributions. Perhaps the oldest and most widely used particle sizing instrument is the optical microscope. Particle dimensions are measured either directly with a "ruler" (*i.e.,* a series of lines of known, equal spacing etched on an insert of the microscope eyepiece) or by comparison with a series of dots or rectangles (also etched on an eyepiece insert) of known size. Numbers of particles in various size ranges are tabulated to yield a size distribution.

The optical microscope method has its advantages and disadvantages in determining particle size distributions. Obviously, the method is slow, tedious and subject to errors due to the small sample size, the few numbers of particles which can be sized in a reasonable length of time and fatigue of the analyst. The major advantage of the method is that one is allowed to literally see the material he is dealing with, and therefore know the morphology of the particles. Particle measuring devices which employ such techniques as light scattering, gravity or electrical resistance to measure particle size do not take particle morphology into account and may, therefore, produce erroneous results when highly irregular shaped particles are measured.

The major disadvantages of the optical microscope method have been eliminated with the advent of automatic image analyzers. Image analyzers allow rapid, reproducible particle size analyses to be performed. Several domestic and foreign manufacturers produce quite a variety of these instruments. Basically, automatic image analyzers consist of two components: a unit for the conversion of the optical image into electrical pulses, and a unit which analyzes the electrical pulses to generate quantitative image information.

The operation of the Imanco Quantimet 720 image analyzing computer will be described as an example of a typical instrument. Applications of image analyzers in general will also be described, with particular emphasis on particle size analysis in fluid systems.

## THEORETICAL OPERATION

The theoretical operation of the Imanco Quantimet 720 image analyzing computer is similar to most image analyzers. The optical image is projected onto the face of a scanner which converts the image into electrical pulses, on the basis of image gray levels. Where possible, conventional digital logic is employed to translate the pulses into quantitative image data. Analogue techniques, however, are still required in some component operations.

While early image analyzers employed TV scanners, most modern image analyzers utilize specially designed scanners which yield optimum compromise between noise, resolution and speed. The Q-720 scanners consist of either plumbicon or vidicon tubes and scan the 720 line raster comprising a standard frame sequentially, 10.5 times per second.[17] Each scan line is divided into 880 picture points horizontally by an internal clock whose period is defined as 1 picture point.[18] This also is the spacing between the scan lines.

As mentioned above, it is the scanner which converts the optical image into electrical signals. The scanner tube (vidicon or plumbicon) is a vacuum tube (Figure 13) with a semi-conductive film target area at the front and a focusable electron beam filament at the rear. The image is focused on the front side of the semiconductive film. A charge builds up on the film proportional to image brightness and flows to the back side of the film creating a current. The electron beam is focused on the rear side of the film and is made to scan the film by a series of deflection coils. As the beam scans, the film is discharged, giving rise to voltage spikes proportional to the image at each discrete point. Height of the pulse is proportional to the gray value of the object while the width of the pulse is proportional to object size.[19]

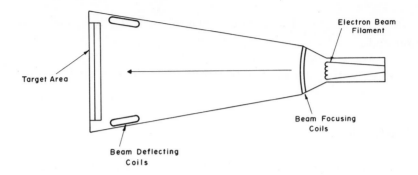

**Figure 13.** The vidicon scanner.

To choose the particular objects, over the background, to be analyzed, a voltage threshold, corresponding to the gray value of the object, must be set. A polarity switch is first set depending on whether white objects on a black background (positive pulses) or black objects on a white background (negative pulses) are to be analyzed. Objects are then "detected" by setting the voltage threshold (with a potentiometer) to allow passage of those signals with a voltage corresponding to the gray level of the desired objects (Figure 14).

With the "detection" properly set, image analysis commences. On each scan line, X number of clock periods are intersected, depending upon the number and width of "detected" voltage pulses present. That is to say, the number of picture points worth of detected objects is determined in each scan line and then stored for processing. Obviously, the number of detected picture points in one scan line will depend on both the number and size (horizontal) of the objects in that line. At a scan rate of 10.5 frames per second, all the data necessary for the system computers to determine such parameters as area, perimeter, intercepts, etc., is available in less than 0.1 second.

From the description of the theoretical operation of the Q-720, it should be obvious that the simplest measurement to obtain is the total area of the detected objects in a field of view; the computer merely adds the number of picture points detected in each of the lines composing one frame. In order to make more complex measurements, such as the number of detected objects present in a field of view, the Q-720 employs

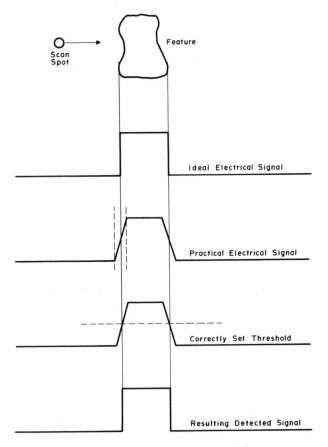

**Figure 14.** Image detection.[20]

a one line "look-back" logic.[18] As a line is scanned, a memory key is set aside at the beam exit point for each object, one or more picture points wide. On the following scan line, the system logic "looks" back to the previous line to determine if a detected object is a new object (no key on the previous line) or is a continuation of an object (key present on the previous line). The memory key on the last scan line intersecting a detected object contains not only the total picture point information for that particular object on that particular scan line, it also contains the summation of the key information for all the scan lines intersecting the object (Figure 15).

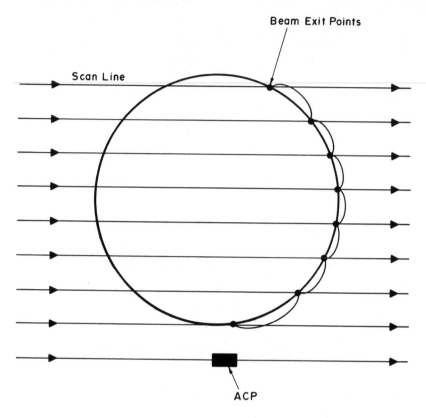

Figure 15. "Keys" containing image information.

## FUNCTIONS AVAILABLE

As with most image analyzers, the Q-720 can provide a variety of image information such as the total projected area, perimeter, longest horizontal and vertical chords (Feret diameter), horizontal and vertical intercepts, and numbers of objects in a field of view (Figure 16). Object size distributions can also be performed using any of the above dimensional parameters. These size distributions can be performed to yield the numbers of objects greater than, less than or in between two stated sizes. In addition, the value of the parameter for a given size class may also be determined; this value, used in conjunction with the number of objects in the chosen size class, yields a mean parameter value for that size class of objects.

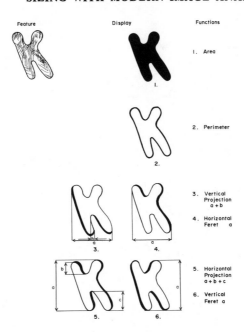

**Figure 16.** Individual functions measured.[5]

With accessory modules, objects may be characterized according to shape. The formula area/(perimeter)$^2$ yields a value specific for individual geometric shapes.[21] This value, known as a form factor, allows separation of objects based on their geometric shape. For example, a size distribution for circles only may be obtained even though there are triangles and rectangles in the same field of view (Figure 17).

Measurements may be carried out on an entire field of view, on a reduced field of view or on individual objects in a field of view.

## APPLICATIONS

The types of samples which can be analyzed by an image analyzer are practically unlimited since most image analyzers offer a variety of image input capabilities. The analyzer may be directly interfaced with the optical and electron microscopes. Alternatively, photographs (and negatives) taken on both types

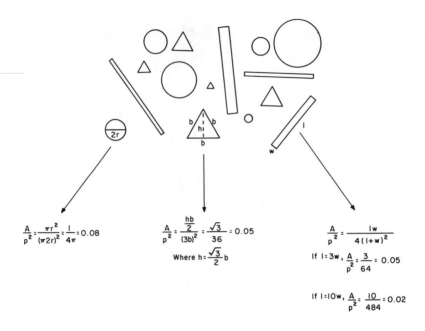

$$\frac{A}{p^2} = \frac{\pi r^2}{(\pi 2 r)^2} = \frac{1}{4\pi} = 0.08$$

$$\frac{A}{p^2} = \frac{\frac{hb}{2}}{(3b)^2} = \frac{\sqrt{3}}{36} = 0.05$$

Where $h = \frac{\sqrt{3}}{2} b$

$$\frac{A}{p^2} = \frac{lw}{4(l+w)^2}$$

If $l = 3w$, $\frac{A}{p^2} = \frac{3}{64} = 0.05$

If $l = 10w$, $\frac{A}{p^2} = \frac{10}{484} = 0.02$

**Figure 17.** Separation of objects according to shape.

of microscopes may be analyzed by imaging the photos on the scanner face through a specially designed light box known as an epidiascope.[22]

The applications of modern image analysis systems are limited only by sample preparation techniques. If sufficient contrast between objects to be measured and the surrounding background can be obtained, a sample can be considered suitable for measurement by an image analyzer. Often, when contrast is insufficient, use of one or more of several available image enhancement techniques can produce a suitable sample. For photographs, contrast may be improved by printing techniques. On the optical microscope, a specialized lighting technique, such as phase contrast, or specialized sample preparation techniques, such as staining (biological specimens) and etching (metallurgical specimens), may produce samples of sufficient contrast for image analysis. The backscatter mode is often used on the electron microscope to make desired objects more visible.

Obviously, the type of image analyzer one has available will also determine the types of samples which can be analyzed. Simple units cannot separate very many gray levels from each other.  A more complex unit is also required to analyze one geometric shape in the presence of others.  One must keep the instrument capabilities in mind when preparing samples for image analysis.

Sample preparation is the key to success in image analysis, especially for particle size analysis.  For dry powder particle size analysis, the powder must first be thoroughly dispersed on a microscope slide in a liquid which provides sufficient particle contrast; the image analyzer will size an agglomerate of two or more particles as one large particle.  Particles in fluid systems must first be removed from the fluid and deposited on a suitable substrate.  This is usually accomplished by vacuum filtering the particle-containing fluid through a membrane filter.  The type of membrane filter employed for particle collection must be chosen to provide maximum particle-to-substrate contrast.  Care must be taken to insure that the suction pulling the liquid through the filter is uniform.  Often the suction around the edges of the filter is stronger than at the center, which results in heavy deposition of particles along the edges; these particles are usually in contact with each other, making image analysis impossible.  Finally, care must be taken to avoid contamination of the particle sample to be analyzed with extraneous debris from the laboratory environment.  This is especially true for filter performance evaluations where the particle size distributions of a test powder are compared upstream and downstream of a filter inserted in a fluid test system.  Contamination of collected particle samples will produce an erroneous evaluation of the tested filter.

## CONCLUSIONS

Modern image analysis systems provide rapid, reliable particle size distributions for a variety of particle types and size ranges. The image analyzer retains the major advantage of the microscope sizing method (*i.e.,* allowing one to physically observe the materials he is dealing with) while eliminating the major disadvantages of slow, tedious and often unreliable data production.

Sample types which can be analyzed by image analyzers are practically unlimited. Bulk samples, such as powders and filters, as well as photographs obtained with an optical or electron microscope, need only have sufficient contrast between objects to be measured and their surroundings to be considered suitable for the image analyzer. Contrast enhancement techniques often yield usable samples from samples with poor initial contrast.

# 5

# SIMPLE SEDIMENTATION METHODS, INCLUDING THE ANDREASON PIPETTE AND THE CAHN SEDIMENTATION BALANCE

Paul C. Siebert
  Pennsylvania State University
  University Park, Pennsylvania 16802

## INTRODUCTION

Sedimentation methods of particle size classification are based on Stokes' Law for the settling velocity of a particle in a fluid. Sedimentation can occur from either a two-layer or a homogeneous suspension system of the particles and the suspending fluid. The Andreason Pipette and Cahn Sedimentation Balance are two commercially available systems that utilize Stokes' Law for particle size analysis from a homogeneous suspension.

## THEORETICAL ANALYSIS

Stokes' Law is derived from a balance of forces acting on a spherical particle in a viscous fluid continuum. In the viscous or Stokes' region, the Reynold's Number based on the particle diameter ($Re_{d_p}$) must be

$$Re_{d_p} = \frac{\rho_p d_p u}{\mu} \leqslant 1.0 \qquad (1)$$

where    $\rho_p$   = particle density

          $d_p$   = particle diameter

          u   = particle velocity relative to the fluid

          $\mu$   = dynamic viscosity of the fluid

In this region, the drag coefficient ($C_D$) is equal to

$$C_D = 24/Re_{d_p} \tag{2}$$

For the fluid to be a continuum with respect to the fluid, the Knudsen Number (Kn) must be

$$Kn = \frac{\lambda}{r_p} < 0.1 \tag{3}$$

where    $\lambda$   = mean free path of the fluid

          $r_p$   = particle radius

Therefore, the viscous or drag force ($F_V$) is by definition of the drag coefficient

$$C_D = \frac{F_V}{\frac{\pi d_p^2}{4} \; 1/2 \; \rho_p u^2} = \frac{24}{(\rho_p d_p u/\mu)} \tag{4}$$

$$\therefore \; F_V = 3\pi d_p \mu u \tag{5}$$

For the "free fall" of a spherical particle from rest in a motionless medium for $Re_{d_p} \leqslant 1.0$ and $Kn < 0.1$, the balance of forces shown in Figure 18 leads to

$$m_p \frac{dv}{dt} = F_G - F_B - F_V \tag{6}$$

Stokes' Law states that the maximum velocity equals the terminal or Stokes' settling velocity ($v_s$) at equilibrium. Thus,

$$m_p \frac{dv_s}{dt} = F_G - F_B - F_V = 0$$

$$m_p g - m_f g - 3\pi d_p \mu v_s = 0$$

$$3\pi d_p \mu v_s = \frac{4}{3}\pi \left[\frac{d_p}{2}\right]^3 g(\rho_p - \rho_f); \; \rho_f = \text{density of fluid}$$

$$v_s = \frac{g}{18} \frac{d_p^2 (\rho_p - \rho_f)}{\mu} \tag{8}$$

which is Stokes' Law. This expression is usable for $1.0\ \mu m \leqslant d_p \leqslant 100.0\ \mu m$ and is most accurate for the range $10.0\ \mu m \leqslant d_p \leqslant 60.0\ \mu m$.

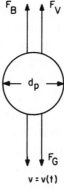

$v = v(t)$

$F_V$ = Viscous Force

$F_B$ = Bouancy Force

$F_G$ = Gravitational Force

**Figure 18.** Force diagram on particle.

This equation can be used for particle size analysis for a known settling velocity of a particle size.

$$v_s = \frac{h}{t} \tag{9}$$

where   h = distance the particle falls in t

      t = time

in the form

$$d_p = \sqrt{\frac{18\ h\mu}{(\rho_p - \rho_f)gt}} \tag{10}$$

Therefore, the mass of particles falling from a height (h) in the time ($t_i$) is the mass of particles of diameter greater than or equal to $d_{p_i}$.

## TWO-LAYER SEDIMENTATION

Two-layer sedimentation involves a very simple theoretical analysis; however, there are practical problems. In two-layer sedimentation the thickness of the suspension layer, $\Delta h$, in which the particles are suspended is small in comparison with the height of fall, h. Once the suspension layer is in place, all of the particles settle through the fluid at their Stokes' settling velocity. Thus, the weight of particles to fall the distance, h, in the time interval $t_i < t \leqslant t_{i+1}$ consists entirely of particles having diameters $d_{p_i} > d_p \geqslant dp_{i+1}$ defined by Equation 10, assuming uniform densities.

The drawback with this approach is that it is very difficult to create a stable suspension layer of high enough concentration ($\sim 1\%$ by volume) for the weights of the size ranges to be accurately measureable.[23] When a layer of suspension of this concentration is floated on a clear fluid, it can be seen that the suspension will rapidly extend into the fluid and turbulently curve downward, especially around the edges of the settling column. This occurs because the density of the suspension is higher than the density of the fluid. Initially, attempts were made to correct this by preparing the suspension in a fluid miscible with but of lighter density than the settling fluid. However, it was found that concentrations of less than 0.05% by volume were required to prevent cluster formation. If clusters form, the measured size distribution is coarser than the actual as the measured settling times will be those of the clusters rather than of the particles. The Micromerograph utilizes two-layer settling through a gas onto a balance pan. The particles are dispersed into the top of the column by a short burst of nitrogen gas. However, Kaye and Weingarten[24] found, by high-speed photography, that the cloud was forced down a seventh of the column before stabilizing. Thus, the real height of fall is less than the column height and the size distribution will again be coarser than actual.

## HOMOGENEOUS SUSPENSION SEDIMENTATION

The sedimentation of a homogeneous suspension involves more complicated theoretical analysis and calculations than does two-layer sedimentation; however, it does not present the practical operating problems. The powder is suspended homogeneously in the settling fluid. Timing of settling begins when stirring of the suspension is stopped. The powder which has settled the distance, h, at any time, t, consists of two fractions:

1. All the particles of Stokes' diameter $d_p > d_p(t)$ as defined by Equation 10.

2. The fraction of particles smaller than $d_p(t)$ that would settle from their actual height in the suspension, $h'$, within the time, t.

Oden's equation states that the percentage weight of particles of diameter greater than $d_p(t)$, $R(t)$, is

$$R(t) = \left[ \frac{W(t)}{W_T} - \frac{dW}{dt} \cdot \frac{t}{W_T} \right] \cdot 100 \qquad (11)$$

where  $W(t)$  = weight of particles falling to height, h, in time, t
       $W_T$  = total weight of particles settled from suspension

or in terms of mass

$$R(t) = \left[ \frac{M(t)}{M_T} - \frac{dM}{dt} \cdot \frac{t}{M_T} \right] \cdot 100 \qquad (12)$$

If the mass falling to the height, h, from the top of the suspension is continuously recorded as with some sedimentation balances, $dM/dt$ is readily available as the tangent of the trace of settled mass vs elapsed time at time, t. Bostock[25] suggests that if the mass on a sedimentation balance is only intermittently recorded, then the data should be plotted as a graph of $M(t)$ vs ln t. This is valid as mathematically $dm/dt \cdot t = dm/d(\ln t)$. However, there is no increased accuracy in this method though the record will be shorter if long sedimentation times are involved.

## SAMPLES OF COMMERCIALLY AVAILABLE SYSTEMS

Numerous systems for particle size analysis utilizing sedimentation by Stokes' Law are commercially available. The majority of these are sedimentation balances by various manufacturers and of varying designs. Two systems, the Andreason pipette and the Cahn sedimentation balance, will be discussed in this paper.

The Andreason pipette (see Figure 19) consists of a sedimentation vessel, a tube for drawing samples, a 10-ml sample section and a two-way stopcock for drawing and draining samples. The sedimentation vessel is filled to a settling height, h, of 20 cm (550 ml) after the suspended sample is added. Sufficient sample is weighed out so that the concentration will be approximately 1% by volume. The pipette is placed in a constant temperature bath both before and after shaking to minimize temperature effects on viscosity of the fluid. Samples of 10 ml are slowly drawn ($\sim$ 20 sec), drained, dried and weighed immediately after shaking and at successive time intervals. The size distribution is calculated from the Stokes diameter $d_p(t_i)$ for each mass fraction for $t_i < t \leqslant t_{i+1}$ as follows:

$$\sqrt{\frac{18\ h\mu}{(\rho_p - \rho_f)gt_{i+1}}} \leqslant d_p < \sqrt{\frac{18\ h\mu}{(\rho_p - \rho_f)gt_i}} \qquad (13)$$

There are several difficulties in the design and application of the Andreason pipette.

1.  The sampling tube interferes with the settling of the particles, especially in the region directly below the tube.

2.  The sampling tube ideally would draw the sample from a thin layer at the 20-cm level; however, the sample is actually drawn from a spherical region around the tube.

3.  A two-layer theoretical analysis is used while homogeneous settling actually occurs.

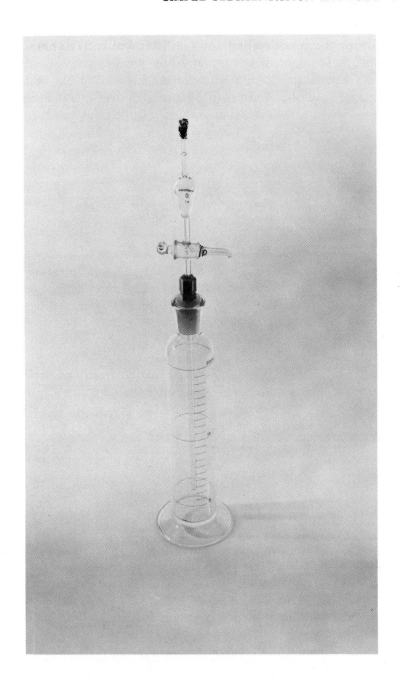

**Figure 19.** Andreason pipette.

Even with these difficulties, the method is correct within 2 to 5% and the equipment is comparatively inexpensive.

The Cahn sedimentation balance (see Figure 20) has many advantages over the Andreason pipette and over many other

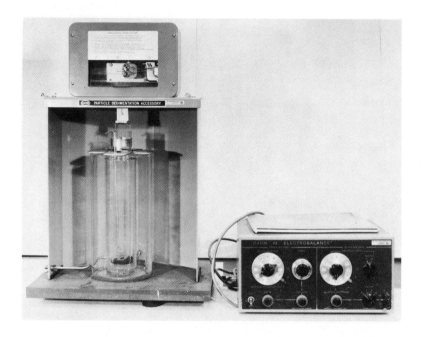

**Figure 20.** Cahn sedimentation balance.

sedimentation balance systems. The system consists of an electrobalance, sedimentation accessory and chart recorder. Samples of 0.1 g to 2.0 g can be used with the system. Some of the desirable features of this sytem include:

1.  The balance pan is suspended from outside the settling column so that settling has no interference.

2.  The balance pan covers the entire bottom of the settling column so that no sample is lost.

3.  The mass of particles settling through the height, h, to the balance pan is continuously and automatically recorded so that the calculations can be made by Equations 10 and 12.

4.  A water bath is provided for temperature control.

5.  A General Electric time-sharing computer program is available for data processing.


Figures 21, 22 and 23 show the precision and accuracy of the Cahn sedimentation balance and compare its resulting size distributions with those of a hydrometer, microscope  and Andreason pipette.

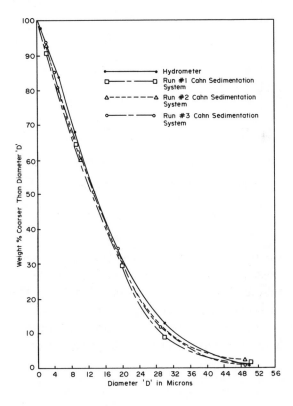

**Figure 21.**  Feldspar.

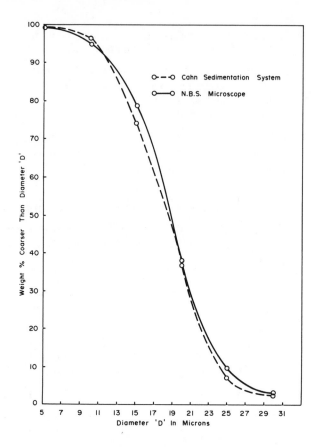

**Figure 22.** Glass beads N.B.S. sample 32.

## SUMMARY

The basic theory of particle size analysis by sedimentation has been discussed. Two commercially available systems using these techniques have been discussed in detail. The Andreason pipette is a commonly accepted, relatively inexpensive system. The Cahn sedimentation balance is more accurate, automatic, and uses a more rigorous analysis; however, it is more expensive.

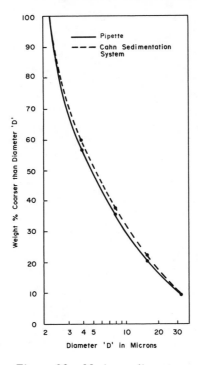

**Figure 23.** Marine sediment.

# 6

# OPTICAL SIZING TECHNIQUES

**Madhav B. Ranade**
IIT Research Institute
10 West 35th Street
Chicago, Illinois 60616

## INTRODUCTION

Methods based on the modulation of light by particles have been employed for obtaining particle size information for a considerable time. Several commercial instruments at various levels of sophistication are available for counting particles, measuring average size and size distributions. Van de Hulst[26] and Kerker[27] have discussed the theoretical aspects of light modulation and have defined the limits of applicability of several techniques useful for particle size measurement. As shown in Figure 24, a light beam incident on the particle is scattered in all directions. The relative angular intensity of light is dependent on the particle size and refractive index, and the wavelength of light. Two main classes of techniques are useful for sizing of particles in a liquid medium: light-scattering and light-extinction.

## LIGHT-SCATTERING

In this technique, light scattered at angles other than 0° may be measured with a photometer. Several useful methods for

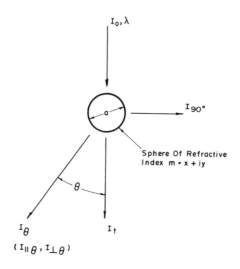

**Figure 24.**   Light-scattering by a single spherical particle.

particle size measurement are listed in Table XII.   Measurements may be made at one angle or several angles in the forward and backward directions.   These methods are based on theoretical light-scattering principles and are applicable in the size ranges indicated in Table XII.   Most methods represent the projected area average particle size.   Methods based on polarization ratio or dissymmetry measurements give particle size directly in a multiparticle system; other techniques provide a measure of projected area of all of the particles and independent knowledge of the number of particles in the viewing volume is required to calculate the size.   Measurement on successive single particles flowing through the viewing volume allows the use of these techniques for size and size distribution measurements.

## LIGHT-EXTINCTION

In the light-extinction technique, the light intensity at 0° or in a small angular range near 0° is measured.   The ratio $(I_t/I_o)$ of the light intensities at 0° with and without the presence of particles is related to particle projected area $(A_p)$ and number concentration (n) by the well known Bouger (or Beer-Lambert) law

Table XII. Optical Sizing Techniques

| Method | Description | Particle Size Range | Particle Size Characteristics |
|---|---|---|---|
| **Light-Scattering:** | | | |
| 1. Angular Variation | Measurement at a continuous spectrum of angles in forward and backward direction | 0.1 to several $\mu$m | Particle size |
| 2. Polarization Ratio | Measurement of two polar components of scattered light at an angle $\theta$ | 0.05 - 1 $\mu$m | Particle size<br>Particle size distribution (several $\lambda$) |
| 3. Dissymmetry Ratio | Measurement of light at two angles $I(90 - \phi)/I(90 + \phi)$ | 0.05 - 1 $\mu$m | Particle size<br>Particle size distribution (several $\lambda$) |
| 4. Near Forward | Measurement of light at an angle (or a range of angles) in the near forward direction | 1 to several $\mu$m | Particle size distribution |
| **Light-Extinction:** | | | |
| 1. Single Particle | Measurement of light at 0° (or a range near 0°) | 1 to several $\mu$m | Average size |
| 2. Transmission (multiple particles) | Measurement of light at 0° (or a range near 0°) | 1 to several $\mu$m | Average size<br>Particle size distribution (several $\lambda$) |

$$\ln \frac{I_t}{I_0} = -K_{ext(a)}A_p nL$$

where: 1. $K_{ext}$ is dependent on the particle size, wavelength of light and the refractive index of the particles;

2. L is the path length of the light beam in the suspension. Independent knowledge of n is required to obtain the average projected particle area $A_p$ which is equal to $\pi a^2$ for a sphere.

## INSTRUMENTS BASED ON OPTICAL MEASUREMENT

Several commercial instruments are available to measure particle size and distribution in various size ranges. A review of several instruments was presented by Davies.[28] A list of commercial instruments is presented in Table XIII.

The Brice-Phoenix photometer is probably the most versatile instrument for particle size measurement. Light intensity over the range of ±135° can be measured by rotating the detector around the sample (Figure 25). The incident light is provided

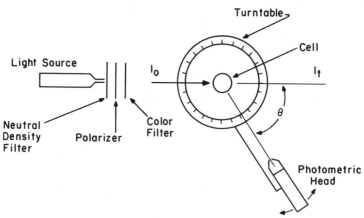

**Figure 25.** Brice-Phoenix photometer.

by a mercury vapor lamp and rendered monochromatic by four different filters. An optional He-Ne laser has been introduced recently. Combinations of slit widths, measuring cells and detector allows all the measurement techniques listed in

Table XIII  Commercial Optical Sizing Instruments

| Instrument | Operating Principle | Particle Size Range | Particle Concentration Range | Manufacturers |
|---|---|---|---|---|
| 1. Brice-Phoenix Universal | Light-scattering and extinction | 0.01 to several $\mu$m | $10^3$ particles/cm$^3$ | Phoenix Precision Inst., Gardner, N.Y. (Div. of Virtiss Co.) C. N. Wood Mfg. Co., Newtown, Pa. |
| 2. Royco Model 345 | Light-extinction/ total scatter | 5 to 400 $\mu$m (five channels) | $10^3$ particles/cm$^3$ | Royco Instrument Co., Menlo Park, Calif. |
| 3. Climet | Near forward scattering | 2 to 200 $\mu$m (two channels) | 1 to $10^8$ particles/cm$^3$ | Climet Instrument Inc., Sunnyvale, Calif. |
| 4. HIAC-SS | Light-obscuration | 2 to 1000 $\mu$m (five channels) | $3.5 \times 10^3$ particles/cm$^3$ | HIAC, Claremont, Calif. |
| 5. Biophysics | Light-extinction, scattering | | | Biophysics, NY |
| No. 6303 | (count only) | | — | |
| No. 6302 | (count and size) | 1 to 100 $\mu$m | — | |
| No. 6303 | (count, size and differentiate) | 1 to 100 $\mu$m | — | |
| 6. Procedyne No. PA-110 | Light-extinction/ scanning | 5 to 1000 $\mu$m | — | Procedyne, New Brunswick, N.J. |

Table XII.  Primarily, this instrument is used with a batch sample.  However, continuous measurement at one angle is possible for continuously flowing suspensions using a special cell.  Continuous dissymmetry measurement is also possible with the optional second detector.  This unit is primarily a research instrument.

The Royco Instruments Company markets the Model 345 unit for particle size analysis and counting of liquidborne particles above 5 $\mu$m in size.  They also provide a lower detection limit of 2 $\mu$m on special orders.  The Model 345 is based on light-absorption technique.  A schematic representation of the optics is shown in Figure 26.  Liquid samples are run at a flow

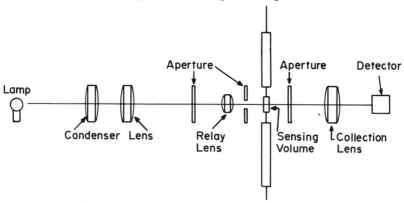

Figure 26.  Schematic diagram of Royco 345.

rate of 100 cm³/min.  Up to 1000 particles/cm³ are handled by the instrument.  Several interchangeable modules permit the data to be displayed and recorded.  Several size ranges beginning at 5 $\mu$m up to 50 $\mu$m and above are standard.  The largest particle handled by the instrument is estimated at 400 $\mu$m.

Climet Instrument Company manufactures a liquidborne particle counter using a near-forward scattered-light measurement.  This unit has two channels of readout, 2 to 20 $\mu$m and 20 to 200 $\mu$m.  It samples at a rate of 200 cm³/min.

HIAC Corporation markets the HIAC-SS counter.  The fluid containing the particles flows through a rectangular passage, past a window, through which light passes to the photodetector.

Pulses generated by the "shadows" of the particles are processed in five channels, each present at a threshold lower size limit. The total size range covered is claimed to be 2 $\mu$m to 1 mm but only part of this wide range may be used at one time. Up to 3500 particles/cm$^3$ are handled by this instrument. A field version of the HIACC-SS is also avilable with two size ranges.

Biophysics Inc. has developed instruments for counting and sizing as well as counting particles (Model 6303 for counting and 6302 for sizing and counting). These units were primarily designed for biological cells. A batch sample passes through a sample cup at a rate of 0.5 ml/min. Each particle crossing the path of a laser beam is detected due to the attenuation of the laser beam.

The Model 6303 simply counts particles. Model 6302 has a larger viewing volume and sizes the particles in the 1 to 100-$\mu$m size range. The data is presented in the form of a 100 channel histogram.

In another model (No. 6301), differentiation between biological cells is provided by staining the cells and measuring light scattered in two angular regions. The cells are detected in the two angular regions depending on the staining and the cell type. The Biophysics instruments are potentially valuable in particle identification; although at present, they are limited to biological systems.

Procedyne Inc. has an on-line particle size analyzer (Model PA-110). The unit is based on a flying spot scanner utilizing a low power He-Ne laser. Particles in the 5 to 1000-$\mu$m size range, in 2% by weight slurry, may be handled at a rate of 5 gal/min. Due to the scanning scheme, particle shape and edge effects are sources of errors in addition to the coincidence (sensing more than one particle as a single unit due to the overlap in the light path). The random chord scan of the laser beam is statistically related to the true projected area diameter. The data is displayed in five channels of readout. The instrument is primarily designed for on-line application but has the capability of use as a batch system.

## COMMENTS

Several commercial optical sizing instruments are available on the market for particles above a few microns. Most of the commercial instruments have been designed to be applicable as rapid response, on-stream size analysis. For smaller particles, only research instruments, such as the Brice-Phoenix photometer, are available. Further developments using light-scattering techniques for smaller particles are possible and are urgently needed.

# 7

# THE "ELECTROZONE" COUNTER: APPLICATIONS TO NONAQUEOUS PARTICLE-FLUID SYSTEMS

Jean Graf

IIT Research Institute
10 West 35th Street
Chicago, Illinois 60616

## INTRODUCTION

The need for particle size distribution data by filter evaluation laboratories is essential. If a filter is introduced into a fluid system to prevent passage of particles large enough to damage the system components, then the performance evaluation of that filter must deal with particle size. A common evaluation method is to determine the change in the particle size distribution of a test material effected by the filter. Particle size distributions must, therefore, be determined both upstream and downstream of the filter in order to calculate removal efficiencies for specific particle size classes.

Particle size analyses can be performed by a variety of techniques and instruments, as is evident by the other discussions in this program. This paper describes the use of electrical sensing zone ("electrozone") instruments such as the Electrozone-Celloscope and the Coulter Counter®. These instruments, which both operate on the same physical principle, provide rapid, automatic particle size data based on particle volumes.

## INSTRUMENT PRINCIPLES

The "electrozone" counter determines the number and volume of particles in an electrically conductive liquid by application of a resistance principle. In application of this principle, a particle suspension flows through a small aperture having an immersed electrode on each side, as shown in Figure 27. The

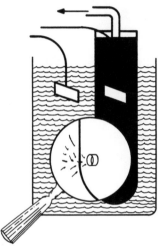

**Figure 27.** Basic mechanism of principle.

electrical resistance to the current applied to the electrodes is determined by the aperture size and the electrolyte strength. As each particle passes through the aperture, it replaces its own volume of electrolyte within the aperture; thus, momentarily changing the resistance value. This change produces a voltage pulse of short duration having a magnitude proportional to particle volume. The resultant series of pulses from the flow of a particle suspension through the aperture is electronically amplified, scaled and counted.

"Electrozone" counters analyze particles between 0.3 and 800 $\mu$m. To cover this broad range, several aperture sizes are needed. For a given electrical condition and aperture size, instrument response is essentially linear with particle volume, provided that the particle-to-aperture diameter ratio is between 0.04 and 0.4. Electronic noise sets the lower detection limit

while the possibility of aperture blockage requires that particles to be sized be below 40% of the aperture diameter.

Research[29] performed at IIT Research Institute has improved
the resolution and extended the potential range of an orifice
by both contouring the aperture and employing a flow-
directional collar fitted with a micro-mesh screen. The schematic of the system is shown in Figure 28. Oversizing errors,

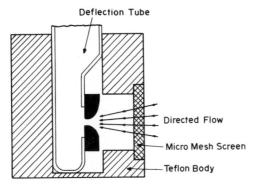

Figure 28. Assembled contoured tube and flow-directional device.

caused by turbulence vortices associated with particles entering
straight-sided (conventional) aperture tubes (Figure 29), are

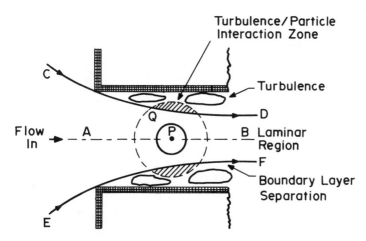

Figure 29. Conception of stream separation.

eliminated by the contoured aperture; the flow-directional collar insures that particles enter the center of the aperture, thereby eliminating another source of turbulence-associated noise. The micro-mesh screen, with openings no greater than 40% of the aperture diameter, eliminates the necessity of "pre-scalping" the particle suspension to prevent aperture blockage.

## SAMPLE PREPARATION

The key to the "electrozone" counter operation and its application to a variety of sample types is *electrolyte selection.* Until recently, the low electrical conductivity of lubricating oils, hydraulic fluids. and other nonaqueous system fluids ruled out "electrozone" counters for routine particle analyses in filter performance evaluations. Several alternative methods of sample preparation have been developed which are particularly applicable to particle samples suspended in nonaqueous solvents.

Nonaqueous electrolytes have been developed for use with oils and water-soluble particles.[30] Jet fuel oil and light mineral oils are miscible to over 40% in the electrolytes listed in Table XIV. Other oils, such as lubricating oils and hydraulic

Table XIV  Nonaqueous Electrolytes[31]

| Salt | Concentration | Solvent |
|------|---------------|---------|
| Ammonium Thiocyanate | 4% | Isopropyl Alcohol |
| Potassium Thiocyanate | 1-2% | Acetone |
| Lithium Iodide | 20% | Isopropyl Alcohol |
| Ammonium Thiocyanate | 4% | Dimethyl Formamide |

oils, are immiscible; such materials require a "coupling solvent" to prevent the oil and electrolyte from separating phases. Table XV lists common oil-coupling solvent-electrolyte systems.

There are several deficiencies with the nonaqueous electrolyte systems. Phase separation can occur even with "coupling

Table XV  "Coupling" Electrolytes[3 1]

| Formula (Volume Portions) | Use |
|---|---|
| 1/3 oil/1/3 ethylene dichloride/1/3 4% NH$_4$CNS in isopropyl alcohol | MIL-H-5606, hydraulic oil, lube oils |
| 1/3 oil/1/6 dimethyl formamide/1/2 4% NH$_4$CNS in isopropyl alcohol | MIL-L-7808, JP-4 Jet fuel |
| 2/5 fuel/2/5 4% NH$_4$CNS in isopropyl alcohol/1/5 ethylene dichloride | Diesel fuel |
| 1/6 oil/1/3 3% NH$_4$CNS in isopropyl alcohol/1/6 trichloroethylene/1/3 tetrahydrofuron | Oronite MLO-8200 and 8515 |

solvents" present; therefore, counts must be made immediately after mixing. The salts used in the electrolyte systems are not soluble in all the system components and may precipitate, giving rise to false counts. Of course, concentration errors in the final data may occur because so many mixing and dilution steps are involved. The most severe fault of the nonaqueous electrolyte scheme is the high level of instrument noise caused by the high electrical resistivities of the nonaqueous electrolytes. The high resistance of the combination electrolytes prohibits use of apertures less than 100 $\mu$m; therefore, the lower particle size limit is only 10 $\mu$m (as opposed to 4 $\mu$m for a typical aqueous electrolyte solution with lower resistance).

Three other sample preparation techniques have been developed for application of "electrozone" counters to filter performance evaluations. The techniques utilize membrane filters to collect the particles to be analyzed from the contaminated fluids. Millipore (cellulosic membrane) filters are generally preferred because of their faster filtering properties. However, Nuclepore (polycarbonate film) filters may be more desirable for some applications. The choice of filter will depend on the type of fluid to be filtered (*i.e.*, fluid viscosity, solvents which will remove all traces of the filtered fluid from the membrane without dissolving it, etc.) and the technique to be used for particle liberation from the filter.

Regardless of the membrane filter employed, particle collection procedures involve identical steps. Aliquots of the upstream and downstream fluids are vacuum filtered through individual preweighed membrane filters. All traces of the contaminated fluid are rinsed from the filters with a suitable solvent before the membrane filters are dried and reweighed. The choice of techniques for liberation of the particles from the membrane filters for "electrozone" counter analysis is dependent on the size and type of particles to be analyzed, as will be evident in the discussions of the three candidate techniques.

A common method for liberation of particles from Millipore membrane filters is to dissolve the membrane. A solution of 5% potassium iodide (KI) in dimethyl formamide (DMF) dissolves Millipore filters and results in an electrolyte with a resistance of 19,000 ohms with a 140-$\mu$m aperture and 30,000 ohms with a 50-$\mu$m aperture.[30] Care must be taken to use extremely clean starting solutions. The KI-DMF solution must be prefiltered through a Nuclepore filter to produce a low-background electrolyte.

The remaining two methods for liberation of particles from membrane filters involve transferring the particles into aqueous electrolyte solutions. The use of aqueous electrolytes offers many obvious advantages: an inexpensive, precleaned, off-the-shelf electrolyte solution (available from Coulter Electronics, Inc.) may be employed thereby eliminating the compatability, toxicity and expense problems associated with nonaqueous electrolyte systems; and extensive technical and operational data is available in the literature for "electrozone" counters operated with aqueous as opposed to nonaqueous electrolytes.

Ultrasonic techniques have been employed to transfer particles directly into aqueous electrolyte solutions. The membrane filter and electrolyte are placed into a beaker immersed in an ultrasonic bath. The type and size of particles present, as well as the type of membrane filter and electrolyte solution used, will dictate the amount of sonication required to adequately remove the particles. In general, Nuclepore filters are best suited for ultrasonic particle removal; the membranes are considerably more resilient than Millipore membranes and less likely to disintegrate in the ultrasonic bath. In addition, the Millipore filter surface is extremely textured—particles can become "buried" in this texture; the Nuclepore surface, on the other hand, is extremely smooth. Research[32] at IITRI has shown that an aqueous sodium pyrophosphate ($Na_4P_2O_7$) solution at 50°C is an especially efficient medium for ultrasonic particle removal. The off-the-shelf sodium chloride solutions for "electrozone" counter analyses may be used, with lower particle removal efficiencies.

However, there are several problems with the ultrasonic particle removal technique which should be investigated before the technique is adopted for a particular type of contaminant particle. Particle removal, especially for small particle sizes, is not 100% efficient. Usually, though, the smaller particles are either too small to be detected by the "electrozone" counter or below the lower cut-off size for the filter being evaluated. The few numbers of larger particles not removed from the membrane filter normally have no statistical effect on the particle size distributions. Care must also be taken to avoid fracture of fragile particle types.

A third sample preparation technique involves liberation of particles by destruction of the membrane filter. The particle-containing membrane filter is almost totally destroyed in an oxygen-plasma, low-temperature asher. When operated according to the manufacturer's instructions,[33] Millipore and Nuclepore filters are reduced to ashes of 0.059%[34] and 0.08%,[35] respectively, of their initial weights. When the particles and residual ash are suspended in aqueous electrolyte with the aid of an ultrasonic bath or probe, the particles of interest deagglomerate and disperse while the ash disintegrates into particles too small to be detected by the "electrozone" counter. In general, the types of contaminant particles of interest to filter evaluation laboratories will be unaffected by the low-temperature asher. Care must be taken, however, to insure that the particles are sufficiently deagglomerated by the ultrasonic disperser without fracturing any of the larger particles.

## DATA ACQUISITION

The simplest procedure for evaluating a filter's performance is to install the filter in a closed fluid test system containing a known amount of a particulate contaminant. Aliquots of the contaminated fluid are withdrawn upstream and downstream of the filter at various time intervals. One of the several methods described in the previous section is then employed to prepare the aliquots for "electrozone" counter particle analysis. Ideally, the fluid test system should contain the type of fluid the filter would encounter during normal operation. In addition, the particulate contaminant should also be representative, both in chemical and physical properties, of the type of particles which challenge the filter in normal operation.

Instrument operating procedures and the format of the data obtained are dependent on the type of "electrozone" counter employed. Models available range from a modest unit which provides cumulative count data for one particle size at a time (*i.e.,* data are given as numbers of particles counted above the manually selected size) to complex, computer-aided systems which generate complete, graphically displayed, particle size distributions.

Most of the instruments available provide a flow metering device which allows particle concentration to be calculated from the number of particles counted in the metered volume of liquid. However, in the case of samples prepared by collection of the particles on a membrane filter, it is advisable to calculate particle concentration by determining the mass of particles deposited on the membrane filter from the known volume of liquid.

The time required to complete an "electrozone" counter analysis is dependent upon several factors; obviously, the type of instrument used is the major factor. If a modest unit is used, several determinations at each of at least four manually selected particle sizes can require up to 20 minutes to perform. If the more complex unit, which measures several particle sizes simultaneously, is employed, sufficient data to produce a statistically valid particle size distribution can be obtained in as little as 5 minutes. The particle suspensions themselves will also determine the amount of time required to complete an analysis: if a wide range of particle sizes is present, two or more apertures will have to be employed; low particle concentrations require extended analysis time. The aperture also affects the speed of analysis by determining the maximum fluid flow rate.

## DATA INTERPRETATION

A variety of statistical methods is avilable for interpreting the particle size data generated by the "electrozone" counter. Typically, the upstream and downstream data are plotted on the same graph for comparison purposes. Straight-line graphs are generally easier to compare and, therefore, particle size data is usually plotted as log of the particle size vs probability (cumulative percent) or log of the number of particles vs log squared of the particle diameter.

The particle size distribution data obtained with the "electrozone" counter may be evaluated in its generated form for most particle types. However, if the particulate test material is composed of highly irregularly shaped particles, the "electrozone" counter must be recalibrated. As stated in a previous section, the "electrozone" counter measures particle volume; what is reported as particle size, however, is the diameter of a sphere having the equivalent measured volume. For flattened, angular particles especially, this equivalent spherical diameter does not truly reflect the size of the particles passing (or being removed by) the tested filter (Figure 30). To compensate for this apparent error, the test particulates should first be sized with the optical microscope so that the true projected diameter particle size distribution may be determined. The test material is then analyzed by the "electrozone" counter, and a correlation is made between the "electrozone" diameter and the microscope diameter.

## SUMMARY

The "electrozone" automatic particle counter may now be applied to particle size analyses in nonaqueous fluid systems. This

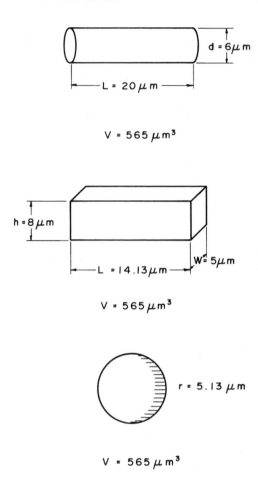

**Figure 30.** Three particle shapes with equal volumes but different dimensions.

is especially significant to filter performance evaluation laboratories since the majority of the particle size analyses, which are
a major part of filter performance tests, are conducted on particles found in nonaqueous fluid systems. Sample preparation
methods involve the use of salts dissolved in nonaqueous solvents which in turn are miscible with the particle suspension
fluid to create electrically conductive media; and removal of
particles from the nonaqueous fluid with a membrane filter
followed by liberation of the removed particles into an aqueous
electrolyte.

A filter performance evaluation should consist of determining particle concentrations and size distributions both upstream and downstream of the filter in a fluid test system; a test particulate of known characteristics is injected upstream of the filter. Time resolved samples present a picture of filter performance as a function of particle loading.

Data obtained must be interpreted carefully.  "Electrozone" counter data obtained with test particles of irregular dimensions will not present a true picture of filter performance unless the counter is calibrated with particle size data obtained with the optical microscope.

# 8

# CENTRIFUGAL PARTICLE SIZE ANALYSIS AND THE JOYCE-LOEBL DISC CENTRIFUGE

Joseph Puretz
    Illinois Institute of Technology
    Technology Center
    Chicago, Illinois 60616

## INTRODUCTION

Simple gravitational settling or sedimentation has long been used to obtain particle size distributions of liquid suspensions. This method is particularly useful when the particle size is broad. The principle of sedimentation is based on the fact that particles falling through a viscous medium quickly attain a terminal velocity and settle at this constant velocity for the remainder of their descent. The equation relating the parameters of the suspending medium and those of the particle is known as Stokes' equation and is given by

$$V = \frac{2gd^2(\rho_1 - \rho_2)}{9\eta} \qquad (14)$$

where   $V$ = the particle's terminal velocity
$g$ = the acceleration due to gravity
$d$ = diameter of equivalent sphere
$\rho_1$ = the density of the particle
$\eta$ = the viscosity of the suspending medium
$\rho_2$ = the density of the suspending medium

To apply this method correctly, it is necessary to keep the concentration of the particles in the fluid sufficiently low to prevent particles from interferring with each other's rate of fall. Customarily, the concentration is kept below 2% by volume. It should be noted that Equation 14 is rigorously valid only for spherical particles but its application to irregularly shaped particles is widely used and shows good internal consistency for randomly oriented falling particles.

In any sedimentation method, including centrifugal, there are essentially two ways of obtaining a size distribution. Either (1) the concentration at a level below the surface is obtained, as in the Andreason pipette, or (2) the overall concentration between the surface and a particular level is obtained. The latter usually involves a more complicated mathematical analysis. Particle sizing based on sedimentation may include any of several different methods applying either of the above techniques. These are:

1.  pipetting or otherwise withdrawing a sample at various times for a given depth or various depths for a given time;

2.  measuring the variation in pressure at various times;[36]

3.  measuring the constant variation of the density of the suspension with time using a hydrometer;[37] and

4.  measuring the change of weight due to particles accumulating on the arm of a balance.[38]

## CENTRIFUGAL SEDIMENTATION

Of the above methods, the one commonly used with centrifugation allows the suspension to "settle" for a predetermined length of time after which a sample is withdrawn. The methods of centrifugal particle size analysis have evolved from the early, primitive techniques, e.g., Romwater and Vandl,[39] who used an ordinary laboratory centrifuge, to those in current use, e.g., the Joyce-Loebl Disc Centrifuge.

Slater and Cohen[40] developed a centrifugal particle size analyzer capable of particle size measurement in the 0.1 $\mu$m to

5 $\mu$m range.   This was essentially a centrifugal Andreason pipette (Figure 31).   The centrifuge vessel was disc-shaped and

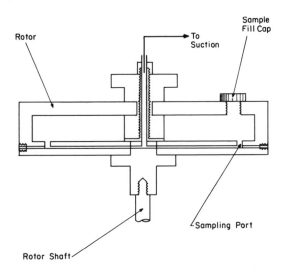

**Figure 31.**   Slater-Cohen disc centrifuge.

rotated horizontally with eight holes in the base evenly spaced on the circumference of a circle.   The holes led into radial tubes which connected to a tube running through the center shaft.   Suction was applied to the shaft and the sample was removed after various times while the centrifuge rotated at a constant speed of 500 rpm.   The fraction was then weighed and a mass distribution was determined.

Burt and Kaye[41] discussed a centrifugal photosedimentometer which they compared to the centrifuge of Slater and Cohen.[40] Good agreement was observed.

## JOYCE-LOEBL DISC CENTRIFUGE

A particularly simple way of obtaining size data of suspensions involves the use of the Joyce-Loebl Disc Centrifuge* (J.L.).   This instrument is capable of sozing particles from 0.01 $\mu$m to 50 $\mu$m, depending on the density.   As the name

---

*Joyce-Loebl Instruments, 20 South Ave., Burlington, Mass.

implies, the Joyce-Loebl Disc Centrifuge uses centrifugal means to achieve particle "settling". This, of course, replaces gravitational acceleration with centrifugal acceleration. Therefore, a modification of Stokes' equation is required. Equation 14 is the unmodified Stokes' equation, which is used in ordinary sedimentation. For the J.L., Equation 14 becomes

$$V = \frac{2a_R d^2 (\rho_1 - \rho_2)}{9\eta} \qquad (15)$$

where $a_R$, the radial acceleration, has replaced g. Substituting $a_R = \omega^2 r$ into Equation 15, we get

$$V = \frac{2\omega^2 r d^2 (\rho_1 - \rho_2)}{9\eta} \qquad (16)$$

In the centrifuge, $V$, is a function of radial position and should be written as $V = \frac{dr}{dt}$, the rate of change of position with time. Substituting into Equation 16 and rearranging before integrating, we have

$$\frac{dr}{r} = \frac{2\omega^2 d^2}{9\eta} (\rho_1 - \rho_2) dt \qquad (17)$$

Integration and rearrangement gives

$$d = \sqrt{\frac{9\eta}{2\omega^2 (\rho_1 - \rho_2) t}} \; \ln \frac{r_2}{r_1} \qquad (18)$$

where   $r_1$ = the position of the particle at time t = 0

$r_2$ = the final position of the particle

t = the length of time required to reach position $r_2$ in seconds

$\omega$ = the rotation speed in radians/second

Equation 18 is the modified Stokes equation which is used in the operation of the J.L.

## Description of the Centrifuge

The basic instrument consists of two parts: (1) the centrifuge unit, comprised of the rotor, the injection head and the sampling head, and (2) the control unit.

### Centrifuge Unit

The centrifuge rotor, Figure 32, is machined from a clear, solid plexiglass-like material, with a 360° annular cavity which

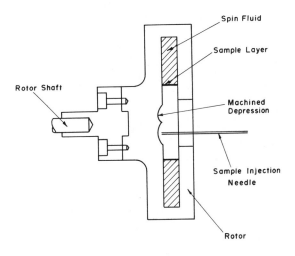

**Figure 32.**   Side view of rotor.

contains the "settling" medium, called the spin fluid, and the suspension being tested. The rotor is fixed vertically and can be operated at several different angular velocities from 1000 to 8000 rpm in increments of 1000 rpm. The suspension is injected at a depression machined into the rear inner surface of the rotor, Figure 32. The injection head is swung into place and the sample is introduced, automatically depressing a switch which begins a preset timing cycle. After injection, the sample accelerates away from the center of the depression and flows in a spiral onto the surface of the spin fluid. A thin-walled, stainless-steel tube samples the suspension. The sampling head has been developed to minimize air turbulence and disturbance of the suspension as the sample is removed. The axis of the sample tube is not concentric with that of the rotor and is positioned in such a manner as to remove the sample in the shortest possible time, Figure 33.

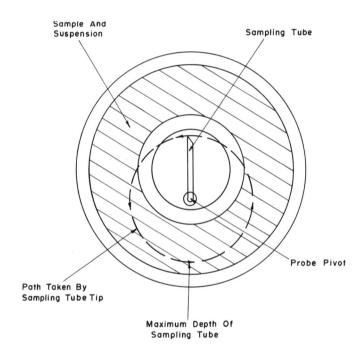

Figure 33. Cross section of rotor.

The undersize fraction is collected and deposited into a volumetric flask. As the tube samples the suspension, it slows down progressively to compensate for the increase in the liquid surface area as more fluid is removed. After completing 180° of travel, the probe depth reaches its maximum final radius, completes the cycle, and returns to its initial position.

*Control Unit*

The electronic control unit consists of the appropriate electronics to accurately control the rotational speed to within ±0.005%. This degree of stability ensures the possibility of accurate mathematical predictions.

The unit also synchronizes a stroboscope with the disc rotation. This feature is indispensable for studying the stability aspects of the suspension as discussed below.

## STABILITY

Stability must be achieved to obtain accurate particle size data. The general stability criterion depends on the fact that a higher density fluid placed onto a lower density fluid produces an unstable density relationship whereby the two fluids try to exchange places. In the centrifuge, this instability is called streaming. If it occurs, streaming will appear in one of two forms: there will appear either (1) tentacle-like arms radiating outward from the interface, or (2) the streaming will occur along the settling path. These difficulties are not merely limitations of the J.L., but could occur in any centrifuging method.

Each form of streaming has its own cause. At the interface, streaming is caused by the absence of streamline conditions of laminar flow due to the relatively high g forces (above 100 g) inside the rotor. The turbulent motion of the particles, together with the viscosity of the spin fluid being greater than the viscosity of the suspension, leads to formation of aggregates and produces a higher mean density than the spin fluid. Together, these effects cause streaming. The practical solution is to add a thickening agent to the sample dispersion, thereby increasing the viscosity without changing the density. Table XVI gives viscosity and density data on various fluids. Streaming along the settling path usually appears when the particles in suspension are close in size. A well-defined particle front is produced which momentarily increases the mean density of this thin layer of fluid. Small streamers appear, injecting particles into the first few layers of the spin fluid, and disappear.

Table XVI. Spin Fluid Data

| Base Spin Fluid | Base Fluid Proportions | Additive | Additive Proportions | Viscosity in Centipoise | Density (g/ml) |
|---|---|---|---|---|---|
| **Stoddard Solvent + Carbon Tetrachloride** | | | | | |
| Stoddard Solvent | 100% by vol | | | 0.621 | 0.86 |
| | | Carbon Tet | 100% by vol | 0.945 | 1.592 |
| Stoddard Solvent | 80% by vol | Carbon Tet | 20% by vol | 0.732 | 0.926 |
| Stoddard Solvent | 60% by vol | Carbon Tet | 40% by vol | 0.760 | 1.093 |
| Stoddard Solvent | 40% by vol | Carbon Tet | 60% by vol | 0.785 | 1.251 |
| | | | | | |
| **Ethanol + Water** | | | | | |
| Water | 100% by wt | | | 1.005 | 1.00 |
| Water | 80% by wt | Ethanol | 20% by wt | 2.183 | 0.97068 |
| Water | 70% by wt | Ethanol | 30% by wt | 2.71 | 0.95382 |
| Water | 60% by wt | Ethanol | 40% by wt | 2.91 | 0.93518 |
| Water | 50% by wt | Ethanol | 50% by wt | 2.87 | 0.91384 |
| Water | 40% by wt | Ethanol | 60% by wt | 2.67 | 0.89113 |
| Water | 20% by wt | Ethanol | 80% by wt | 2.008 | 0.8434 |
| Water | 10% by wt | Ethanol | 90% by wt | 1.20 | 0.78934 |
| | | | | | |
| **Calasec + Water** | | | | | |
| Water | 100 ml | Calasec | 1.0 g | 17.95 | 01.01 |
| Water | 100 ml | Calasec | 1.5 g | 25.40 | 01.015 |
| Water | 100 ml | Calasec | 2.0 g | 36.35 | 01.02 |
| Water | 100 ml | Calasec | 2.5 g | 44.7 | 01.025 |
| Water | 100 ml | Calasec | 3.0 g | 54.5 | 01.03 |
| Water | 100 ml | Calasec | 3.5 g | 64.95 | 01.035 |
| Water | 100 ml | Calasec | 4.0 g | 75.71 | 01.04 |
| | | | | | |
| **Sucrose + Water** | | | | | |
| Water | 100% by wt | | | 1.005 | 1.00 |
| Water | 96% by wt | Sucrose | 4% by wt | 1.15 | 1.014 |
| Water | 90% by wt | Sucrose | 10% by wt | 1.37 | 1.038 |
| Water | 80% by wt | Sucrose | 20% by wt | 1.96 | 1.081 |
| Water | 70% by wt | Sucrose | 30% by wt | 3.20 | 1.127 |
| Water | 60% by wt | Sucrose | 40% by wt | 6.20 | 1.176 |
| Water | 50% by wt | Sucrose | 50% by wt | 15.25 | 1.230 |
| Water | 40% by wt | Sucrose | 60% by wt | 56.50 | 1.287 |
| | | | | | |
| **Glycerol + Water** | | | | | |
| Water | 100% by wt | | | 1.005 | 1.00 |
| Water | 90% by wt | Glycerol | 10% by wt | 1.31 | 1.0217 |
| Water | 80% by wt | Glycerol | 20% by wt | 1.75 | 1.0461 |
| Water | 70% by wt | Glycerol | 30% by wt | 2.51 | 1.072 |
| Water | 60% by wt | Glycerol | 40% by wt | 3.73 | 1.0989 |
| Water | 50% by wt | Glycerol | 50% by wt | 6.05 | 1.1258 |
| Water | 40% by wt | Glycerol | 60% by wt | 11.50 | 1.1528 |
| Water | 30% by wt | Glycerol | 70% by wt | 23.50 | 1.1797 |
| Water | 20% by wt | Glycerol | 80% by wt | 62.0 | 1.2066 |

The solution is to increase the spin fluid density to a point where the introduction of particles into the spin fluid does not cause streaming. In either case, streaming is not a consistent effect and cannot be accounted for in practice but instead must be eliminated entirely to ensure a valid run.

## DESCRIPTION OF OPERATION

The actual operation of the J.L. is straightforward. The parameters to consider are (1) speed of disc rotation, (2) "settling" time, (3) choice of spin fluid  and (4) volume of spin fluid used.

To obtain a size distribution by the J.L., the particle density and the density and viscosity of the spin fluid must be known. If the particle density is unknown, it should be determined by pycnometer. A decision should be made on the number and range of the size fractions of interest. The volume of spin fluid establishes the initial position of the particle, while the final position of the particle is fixed by the endpoint of the sampling tube, Figure 33, and correspond to 5 ml remaining in the rotor. Equation 18 can then be used to calculate the "settling" time for a given particle size. This procedure is repeated for each size fraction of interest until the entire distribution is obtained.

To obtain a set of operating conditions, the following guidelines may be used:

1. **Dispersion.** The dispersion should be well dispersed with a particle concentration between 1/2% to 2% by volume. The choice of concentration should be compatible with the analysis technique, *i.e.,* use enough material to obtain analyzable fractions. A blender or ultrasonic bath may be used to obtain suitable dispersions.

2. **Spin Fluid.** With density and viscosity data on various spin fluids, a suitable choice can be made to achieve reasonable centrifuging times. Of course, a spin fluid should be chosen that is chemically inert to the rotor.

## BRIEF DESCRIPTION OF A RUN

The rotor is brought up to the chosen rotation speed and a predetermined volume of spin fluid between 10 and 40 ml is introduced via a large syringe. The spin fluid is allowed to stabilize for approximately 20 to 30 seconds and usually 1 ml of the sample dispersion is injected with the injection head. The timing cycle begins automatically and at the appropriate time, determined by the modified Stokes equation, the sampling head removes the undersize fraction which is collected. The rotor gradually reduces speed and comes to a stop, containing the oversize fraction which can also be analyzed. This procedure is repeated for each fraction until a distribution is obtained.

## ANALYSIS

Virtually any method of analysis may be used. Table XVII indicates a few of the myriad types of analyses available.

Table XVII. Possible Analytical Methods

| Suspension | Size Range ($\mu$m) | Analysis |
|---|---|---|
| Calcium Carbonate | 0.05 to 15 | Chemical titration by acids |
| PVC Latex Emulsion | 0.1 to 2.0 | Gravimetric |
| Barium Chromate | 0.1 to 4.0 | Colorimetric |
| Blue Pigment | 0.05 to 4.0 | Spectrophotometric |

A new Joyce-Loebl instrument has all of the discussed features in addition to a light source and photodiode combination which enables the J.L. to be used as a photosedimentometer. The output of the photodiode is recorded on a chart recorder, and the particle size distribution is obtained from data reduction of the chart record.

## SUMMARY

The advantages of the J.L. over other centrifugal means of particle size analysis result from (1) the way in which the sample is introduced, (2) the manner in which a sample is removed with a minimum of turbulence and disturbance, and (3) after sample injection, the operation is virtually automatic; with sampling completed, the instrument shuts itself off.

These features give the J.L. method of particle size analysis the following desirable characteristics:

1.   Accuracy to less than 2% uncertainty of cumulative weight undersize

2.   Speed of operation; a complete size analysis can usually be obtained in less than one hour

3.   A wide range of particle size analyses may be handled.

The use of the J.L. to obtain particle size distributions is relatively easy, involving several short algebraic calculations. (Specifications and applications of the J.L. are listed in Table XVIII.)

**Table XVIII.   Specifications and Applications.**

| | |
|---|---|
| Particle Size Range | 0.01 $\mu$m to 50 $\mu$m, depending on density |
| Applications | Virtually any particle may be measured, such as: |
| | Dyestuffs |
| | Pigments |
| | Cements |
| | Ceramics |
| | Emulsions |
| | Pharmaceuticals |
| | Minerals |
| | Metal Powders |
| | Biological Material |
| | Clays and Soils |
| Accuracy | Fractions removed with 98-99% collection efficiency |
| Stability | ±0.005% rotational and ±0.005% in speed |
| Sample Volume | 1/2 ml to 1 ml |
| Sample Concentration | 1/2% to 2-1/2% by volume |
| Fill Volume | 10-40 ml |

# 9

# INERTIAL ANALYZERS FOR PARTICLE SIZE
# MEASUREMENT OF AEROSOLS

Donald L. Fenton
   IIT Research Institute
   10 West 35th Street
   Chicago, Illinois 60616

## INTRODUCTION

The first cascade impactor was developed immediately after
the Second World War. Although detailed knowledge of the
internal flow fields was largely unknown at the time, the impactor received widespread use. The cascade impactor is extremely convenient for obtaining particle size distributions of
an aerosol when the size range falls within 0.5 to 20 $\mu$m.

Basically, an impactor consists of a nozzle and a flat plate
located near the nozzle exit as shown schematically in Figure
34. One nozzle and flat collection plate is called a single stage.
Several impaction stages connected in series are commonly
referred to as a cascade impactor. All impactors utilize either
rectangular or circular cross section nozzles. When operated,
air is drawn through the nozzle, thus generating a high velocity
jet. Near the flat plate, the flow sharply turns and moves
parallel to the plate. Particles, because of their inertia, may or
may not strike the collection plate. Sufficiently large particles
will impact onto the plate rather than flow along with the air
stream. Particle impaction is determined by the balance between
inertia forces and the aerodynamic drag force.

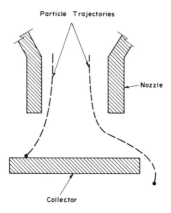

**Figure 34.**  Particle trajectories in an impactor.

An important consideration in the operation of an impactor is the adherence of the particles to the collection plate.  In practice, adherence can be improved through coating the collection surface with a sticky material or substituting a filter (roughened surface) for the plate.  The number of microorganisms per unit volume of air can be determined using a layer of culture medium on the collection plate.

In a cascade impactor, the stages are arranged to permit the jet velocity to increase with each succeeding stage, and therefore cause   the particles of progressively smaller sizes to be impacted.  Air velocities within the nozzle are typically on the order of 100 to 200 m/sec.  Hence, the cascade impactor classifies particles according to their aerodynamic size.  The aerodynamic size is a complicated function of the particle's physical size, shape  and material density.  The aerodynamic size is important because it controls the motion of a particle in an air stream.  Consequently, aerodynamic size is significant for studies concerning lung inhalation, spray effectiveness  and gaseous cleaning devices.

Another sampling instrument, referred to as an impinger, utilizes inertial impaction but deposition occurs at the bottom of a fluid-containing vessel.  The downward-directed air jet displaces the fluid and uncovers the bottom of the vessel.

The particles that impinge against the wet surface are subsequently washed off by the fluid.  The undeposited particles may be caught as the air bubbles rise through the fluid.  The particles are usually examined in the liquid suspension.  Water is the most commonly used fluid.

## BASIC THEORY

A typical collection efficiency curve for a single impactor stage is plotted as a function of the square root of the Stokes number ($N_{st}$) in Figure 35.  The collection efficiency, $\eta$, for

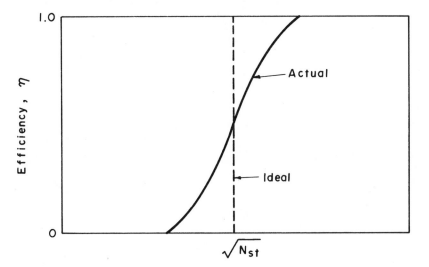

**Figure 35.**  General shape of a typical experimental efficiency curve.

particles of a specific size is defined as the number fraction of the particles which actually impact.  The Stokes number is a dimensionless ratio equal to the particle's stopping distance divided by a characteristic dimension of the system.  The Stokes number is

$$N_{st} = \frac{\rho_p C V_o d_p^2}{9\mu W} \qquad (19)$$

where  $\rho_p$  =      particle material density

        C  =      Cunningham slip correction

       $V_o$  =     average velocity of jet

       $d_p$  =     particle diameter

       $\mu$  =     dynamic fluid viscosity

       W  =     jet width or diameter

An impactor should, ideally, collect all particles larger than a certain size and no particles of smaller size. A cascade impactor with ideal stages would greatly improve the measurement of particle size distributions. Real effects, however, modify the ideal straight line to a "S" shaped curve for each stage.

In general, the size of an irregularly shaped particle may be defined as the diameter of a sphere that is equivalent to the particle with respect to some property readily measured. If this property is of interest, then the defined "equivalent" diameter is merely a convenient form in which to express the amount of that property associated with the particle. The aerodynamic diameter, which is actually a measure of the particle's terminal settling velocity, is an equivalent diameter of this type. Consequently, the aerodynamic diameter is a common deposition parameter and is defined as the diameter of a sphere of unit density having the given value of the settling velocity. The terminal settling velocity is given by

$$V_G = \frac{\rho_p C d_p^2 g}{18\mu} \qquad (20)$$

where g is the acceleration due to gravity. The settling velocity is also used to define another equivalent diameter—the Stokes diameter. The Stokes diameter, $d_s$, is the diameter of a sphere having the same bulk density and the same terminal settling velocity.

An estimate of a particle's aerodynamic diameter ($d_a$) is possible from information concerning the material density of the particle and the measured diameter, d. This is achieved by application of a volume shape factor, $\alpha$, and a resistance shape factor, $\beta$, both related to the measured diameter. Therefore, the aerodynamic diameter is

$$d_a = \left( \frac{6}{\pi} \frac{\rho_p}{\rho_{p_0}} \frac{\alpha}{\beta} \right)^{\frac{1}{2}} d \qquad (21)$$

where $\rho_{p_0}$ represents unit density. After the ratio, $\alpha/\beta$ is determined experimentally for a specific measured diameter and for particles of a given material, it can be applied to similar measurements of diameter, made on particles of the same type, to estimate their aerodynamic diameters.

The first study on cascade impactors was performed by May[42] and since then, a multitude of impactors have been constructed, tested and reported in the literature. Theoretical analysis of impactors was initiated by Baurmash et al. (1949) and is reported by Wilcox.[43] The approach taken was to assume that the flow leaving the jet makes a right angle turn at the impaction plate. A relationship was then derived relating the impaction collection efficiency to the particle's Stokes number.

Davies and Aylward[44] applied the method of conformal mapping to obtain an analytical solution to Euler's equation for flow of a frictionless fluid through a rectangular impactor. The resulting efficiency curves were accurate for impactors operated at only high Reynolds numbers ($N_R$).

Ranz and Wong[45] developed an approximate flow field to determine the efficiencies of both rectangular and round jet impactors. The flow field utilized was a frictionless stagnation flow in the vicinity of the stagnation point. The effect of jet-to-plate distance on the characteristics of the impactor was not taken into account with this solution.

Mercer and Chow[46] modified the Ranz and Wong[45] flow field in a way to account for the variable jet-to-plate distance. Both rectangular and circular jet impactors were investigated.

Marple and Liu[47] solved the governing Navier-Stokes equations by application of finite difference methods. Calculated impactor efficiency curves were generated for the rectangular and circular jet geometries. The effects of jet-to-plate distance, jet throat length and jet Reynolds number were all investigated.

For the two dimensional rectangular jet, the governing Navier-Stokes equations are
x-direction:

$$\rho \left[ u\frac{\partial u}{\partial x} + v\frac{\partial u}{\partial y} \right] = - \frac{\partial P}{\partial x} + \mu \left[ \frac{\partial^2 u}{\partial x^2} + \frac{\partial^2 u}{\partial y^2} \right] \quad (22)$$

y-direction:

$$\rho \left[ u\frac{\partial v}{\partial x} + v\frac{\partial v}{\partial y} \right] = - \frac{\partial P}{\partial y} + \mu \left[ \frac{\partial^2 v}{\partial x^2} + \frac{\partial^2 v}{\partial y^2} \right] \quad (23)$$

and the continuity equation is

$$\frac{\partial u}{\partial x} + \frac{\partial v}{\partial y} = 0 \quad (24)$$

where the x-direction is the transverse distance from the jet axis, y-direction the longitudinal distance from the entrance and u and v the respective components of velocity. Solutions were obtained by Marple and Liu for laminar flow conditions throughout the flow field and constant fluid properties. The boundary conditions included: solid boundaries impervious to flow, no slip flow at the boundaries, symmetry at jet axis and parallel flow at the impaction plate. Governing equations of similar form can be written for viscous flow in a circular jet. Similar assumptions and boundary conditions also apply.

Calculation of particle trajectories is based on the predetermined flow fields. For a rectangular impactor, the equations of motion of a particle are

$$m_p \frac{d^2 x}{dt^2} = \frac{3\pi\mu d_p}{C} \left( u - \frac{dx}{dt} \right) \quad (25)$$

$$m_p \frac{d^2 y}{dt^2} = \frac{3\pi\mu d_p}{C} \left( v - \frac{dy}{dt} \right) \quad (26)$$

where $m_p$ is the particle mass and t time. Marple and Liu[47] have obtained solutions for Equations (25) and (26) and for the case of the circular jet impactor. Experimental verification was obtained for the solutions obtained.

The efficiency curves in Figures 36, 37 and 38 demonstrate the trends established by Marple's work.  Figure 36 shows the

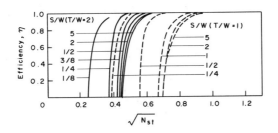

**Figure 36.** Impactor efficiency curves showing effect of jet-to-plate distance for $N_R$ = 3000.[4 7] ——— round impactor; - - - - - rectangular impactor.

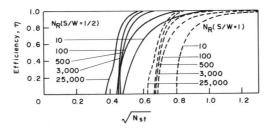

**Figure 37.** Impactor efficiency curves showing effect of the jet Reynolds number.[4 7] ——— round impactor; - - - - - rectangular impactor.

influence of the jet-to-plate distance to jet-width (S/W) for two throat length to jet-width ratios and a constant Reynolds number of 3000.  Figure 37 demonstrates the effect of varying Reynolds number and Figure 38, the effect of jet throat length. From these figures, the following conclusions result:

- rectangular impactors require greater Stokes numbers for particle impaction than circular impactors at the same jet-to-plate spacing

- the slope of the impaction efficiency curve is similar for both rectangular and circular impactors

- the jet-to-plate distance is only important at relatively small values.

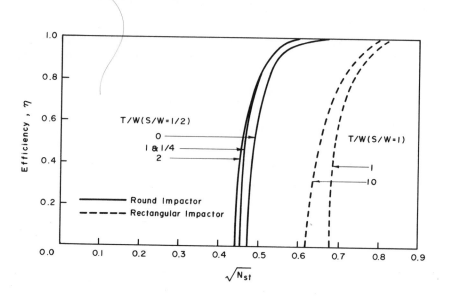

**Figure 38.** Impactor efficiency curves showing effect of the throat length for $N_R = 3000$.[47]

The effects of the Reynolds number on the 50% cut point is shown in Figure 39. The value of $(N_{st_{0.50}})^{1/2}$ is seen to be relatively constant for Reynolds number greater than 100. However, from Figure 37, Reynolds numbers below 500 have a significant effect on the sharpness of cut of the efficiency curves. With this information, impactors can be operated at

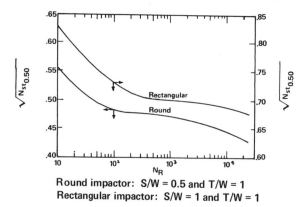

Round impactor: S/W = 0.5 and T/W = 1
Rectangular impactor: S/W = 1 and T/W = 1

**Figure 39.** Effect of Reynolds number of 50% cut point of impactors.

Reynolds number values below 500 if sharp cutoff characteristics are unimportant.  This is only true when S/W ratios are
$\geqslant$ ½ and 1 for the circular and rectangular jets respectively.
Note also the large changes in $(N_{st0.50})^{1/2}$ for Reynolds numbers below 100.  Consequently, a significant error could result
at low Reynolds numbers by calculating a new cutoff particle diameter using data obtained at a different jet velocity.

## CALIBRATION METHODS

In the past, calibration of the impactor before use was
thought necessary.  A test aerosol was sampled by the impactor to determine the mean particle size deposited on each
stage.  Subsequent microscopic examination was used to determine some linear dimension related to the particle size.  The
stage "constants" were then defined as the mass median diameters of the size distributions found on the various stages.
For any given stage, half the mass fraction collected on the
stage plus the cumulative mass fraction collected on previous
stages was plotted as the mass fraction associated with
particles larger than the stage mass median diameter.

Presently, impactors are calibrated using spherical particles
of known density to obtain curves of collection efficiency as
a function of aerodynamic diameter for each impaction stage.
The diameter corresponding to a collection efficiency of 0.5
is assumed as an effective cutoff aerodynamic diameter.  The
assumption is also made that all particles collected at a given
stage, irrespective of shape and density, have aerodynamic
diameters larger than the cutoff diameter for that stage.
Validity of the relationship between collection efficiency and
aerodynamic diameters for spheres having densities approaching
19 $g/cm^3$ is demonstrated by Couchman.[48]   Laskin's data[49]
further support the relationship with irregularly shaped $UO_2$
particles.

The effective cutoff method implies that the efficiency curves
can be approximated by step functions—one step for each stage
occurring at the effective cutoff aerodynamic diameter.  Errors
resultant from this method of calibration originate by particles

not being collected in this manner. A portion of the particles larger than the effective cutoff aerodynamic diameter are not collected and particles smaller than this cutoff diameter are collected. If the two deviating mass fractions distributed about the effective cutoff diameter are equal, then the true effective cutoff diameter equals the effective cutoff aerodynamic diameter. In most cases, the true effective cutoff diameter differs slightly from the effective cutoff aerodynamic diameter. Rectangular jet impactors usually have larger errors of this type than the circular jet impactors.

## OPERATIONAL ERRORS

The separation of different fractions within the impactor is important. Measurements performed by Lippman[50] with $U_3O_8$ particles are shown in Figure 40 for varying mass concentrations. Obviously, the separation is far from complete,

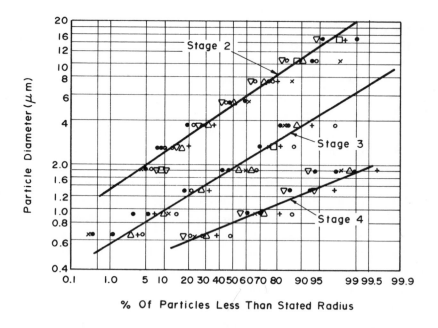

**Figure 40.** Deposition of $U_3O_8$ particles in a cascade impactor.[50]

but the mean diameter and mass of the deposit on each stage are sufficient for most practical purposes.

Errors result due to the operation of the impactor itself. Changes in the aerosol input occur by wall losses and disaggregation. Particle rebounding and reentrainment modify the collection efficiency characteristics.

Nonlaminar flow conditions between the impactor stages encourage wall deposition. The Andersen impactor[51] has negligible wall losses even though the flow makes several right angle turns. The Lovelace impactor[46] is deigned to avoid abrupt changes in flow direction. Wall losses as high as 20% of the total sample have been reported in the literature.

Disaggregation occurs within the jet and is due to particulate relaxation effects (particle and fluid at different velocities). Coal particles have exhibited marked disaggregation when passed through impactors attaining a maximum gaseous velocity of 40 m/sec$^{-1}$.[52] Increasing flow rate also increases disaggregation for a given particulate material. Cascade impactors, because the early stages operate at low flow rates, tend to reduce disaggregation.

Particle reentrainment can be a significant source of error. With rising Stokes number, the collection efficiency will first increase, attain a maximum value, and then decrease, depending on the magnitude of reentrainment. Special coatings and surface roughing are used to minimize this source of error. However, particles below a certain maximum size, dependent on the aerosol, are not likely to reentrain because of surface forces and other effects. Reentrainment also limits the maximum quantity of particulate that can deposit onto a stage, thus providing a "loading capacity." Although the loading capacity itself is not reliable, comparisons among different impactors can be made on this basis.

## TYPICAL IMPACTOR OPERATIONAL DATA

Ranges of performance characteristics and specifications are indicated below for cascade impactors:

| | |
|---|---|
| Number of impaction stages | 1 to 7 |
| Jet geometry | rectangular or round |
| Operating flow rate ($\ell$ min $^{-1}$) | 0.05 to 100 |
| Orifice dimension (cm) | 0.01 to 0.2 |
| Jet velocity (m sec $^{-1}$) | 0.1 to 150 |
| Effective cutoff aerodynamic diameters ($\mu$m) | 0.5 to 20 |
| Jet Reynolds number | 100 to 5,000 |
| Loading capacity (mg) | $0.05 \times 10^{-5}$ to 100 |

# PARTICLE SIZE ANALYSIS BY ELUTRIATION AND CENTRIFUGATION AS EXEMPLIFIED BY THE BAHCO MICROPARTICLE CLASSIFIER*

Hubert Ashley
   IIT Research Institute
   10 West 35th Street
   Chicago, Illinois 60616

## INTRODUCTION

The Bahco, Figure 41, is a versatile particle classifier for powders, dusts and other finely-divided solid materials. A list of materials suitable for particle size analysis by the Bahco is given in Table XIX. The Bahco's working range is from approximately 1 to 60 $\mu$m. Developed in the 1950s, the Bahco has lost much of its initial appeal to the more recently developed techniques (such as electrical-resistant zone counters and sonic sifters) that are faster, provide sharper size grading and require less sample for analysis. However, it retains four important features not found in other classifiers. These are:

- Provides information on the aerodynamic size of particles. This information can be translated into terminal settling velocity useful for designing emission control devices and particle collectors. The Bahco is the recommended method

---

*Available through the H. W. Dietert Co., 9330 Roselawn Avenue, Detroit, Michigan 48204

**Figure 41.** The Bahco Microparticle Classifier.

for the determination of terminal velocity distribution in the ASME Powder Test Code #28.[53]

• Fractionates sample sizes in sufficient quantity for research or further analysis and study.

• Classifies material in the dry state. Thus, solubility problems are avoided.

• Analyzes by simple, reliable gravimetric weighing.

**Table XIX. Typical Materials Sized by the Bahco Microparticle Classifier.**

| | |
|---|---|
| Pulverized coal | Cements |
| Paint pigments | Wood flour |
| Milled products | Soap powders |
| Blood meal | Resin powders |
| Mineral flours | Sintered flours |
| Cosmetics | Abrasives |
| Food powders | Ores |
| Metal powders | Combustion dusts |
| Industrial dusts | Fly ash |
| Chemicals | Clay particles |
| Investment materials | Drugs |

Some of the drawbacks to the Bahco are:

- It takes several hours to complete an eight fraction analysis.
- Gram quantities of material are required for analysis, the minimum quantity needed for a reliable test being 5 g. However, 10- to 20-g quantities are recommended.
- Care must be exercised with certain materials, especially those that are friable or hygroscopic.

## PRINCIPLE OF OPERATION

The Bahco Microparticle Classifier uses a combination of elutriation and centrifugation to separate particles in an air stream.  Briefly, the sample is introduced into a spiral-shaped air current flowing towards the center.  The spiral current of air has suitable values of tangential and radial velocities so that a certain part of the sample is accelerated by the centrifugal force toward the periphery of the whirl, the other part of the sample being carried by the air current toward  the center of the whirl by means of friction between the air and the dust particle.  The size, shape and density of the particle determines which direction it will take in the air current.  By varying the flow, it is possible to change the size cut and, thus, the material can be divided into a number of fractions.

As described in the manufacturer's literature[54,55] the sample passes from the hopper (A), Figure 42, down the feed nozzle (F) where it is carried by an extremely small quantity of air at high velocity into the feed hole.  In the small rotary duct, the air velocity is rapidly reduced as the radius increases and the movement of the material is taken over by the centrifugal force.  During the passage through the narrow duct, where the rotation of the air is high owing to the friction against the walls, the particles are also given a preliminary velocity in the direction of rotation of the air in the sifting chamber (J), to which the sample is introduced through the opening (I).

The spiral of air operating in the sifting chamber (J) is created by the vanes of the fan wheel in the upper rotor assembly.  These vanes draw air through the annular opening between the throttle nut (M) and the lower edge of the rotor

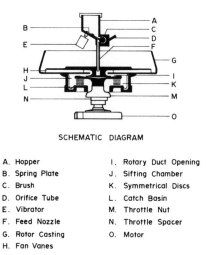

SCHEMATIC DIAGRAM

A. Hopper
B. Spring Plate
C. Brush
D. Orifice Tube
E. Vibrator
F. Feed Nozzle
G. Rotor Casting
H. Fan Vanes

I. Rotary Duct Opening
J. Sifting Chamber
K. Symmetrical Discs
L. Catch Basin
M. Throttle Nut
N. Throttle Spacer
O. Motor

**Figure 42.** Schematic diagram of Bahco classifier.

assembly casting. The position of the throttle nut, as set by the throttle spacer (N), produces a predetermined mass flow of air which is easily altered by the exchange of spacers.

After entering the annular opening, the air passes through a bank of closely spaced discs (K) which impart the necessary symmetrical rotation to the air by the friction against the discs. The air then passes in a spiral through the sifting chamber (J) and is driven out through the vaned fan wheel. The design is such that a homogeneous gyratory field is established in the sifting chamber with streamlines having the same inclination towards the radius and the same velocity at every point of an arbitrary circle within the sifting chamber.

The sample enters the sifting chamber symmetrically from the duct mentioned above and is divided into two fractions: the lighter fraction is borne by the spiral of air out through the fan to the stop-ring section of the rotor casting (G) where it is deposited due to the strong centrifugal force and the relatively low radial air velocity at this point, and the heavier fraction is carried by centrifugal force out against the rotating wall of the sifting chamber where it forms a deposit in the catch basin (L).

As all the surfaces have approximately the velocity of the air, the treatment of the material is gentle so that it is not worn during its repeated passages through the apparatus.

Some of the components of the Bahco can be seen in the disassembled photograph of Figure 43.

Size distributions are obtained by changing the mass flow of air through the sifting chamber. The change is accomplished by the U-shaped throttle spacers inserted between the throttle nut and the rotor assembly casting. Seven spacers of varying thicknesses are generally provided with the instrument. Thus, a nine-size-fraction analysis is possible.

Good results require that the powder sample be deagglomerated and well-dispersed upon entering the sifting chamber. The Bahco provides several features to assist deagglomeration and dispersal and control the powder feed rate. These features are associated with the hopper (see Figure 42) and consist of a spring plate (B), rotating brush (C), orifice tube (D), and vibrator (E). The brush is the primary mechanism for dispersing the powder. As powder feeds from the bottom of the hopper, it is gathered by the brush and flung into a gas stream entering through the orifice tube as the bristles snap past the spring plate. When properly adjusted, the spring plate is in contact with the brush. The gas stream entering the orifice should be either clean, dry compressed air or nitrogen. The proper gas flow is obtained when the pressure drop across the orifice is 1/2 in. of Hg.

The feed rate is controlled by the vibrator and the spring plate. Both are adjustable: the vibrator by a rheostat and the spring plate by a slot and screw assembly. A feed rate of 1 to 2 g per minute is satisfactory.

Proper operation of the Bahco is assisted by drying and sieving the powder prior to placing it in the hopper. Drying at 105 to 110°C to constant weight is recommended. Sieving the dried powder to remove oversize material is also recommended. The manufacturer suggests removing particles greater than 60 $\mu$m while PTC #28 suggests screening through a 100 mesh screen (149 $\mu$m). In our experience, operation is improved by removing particles over 60 $\mu$m.

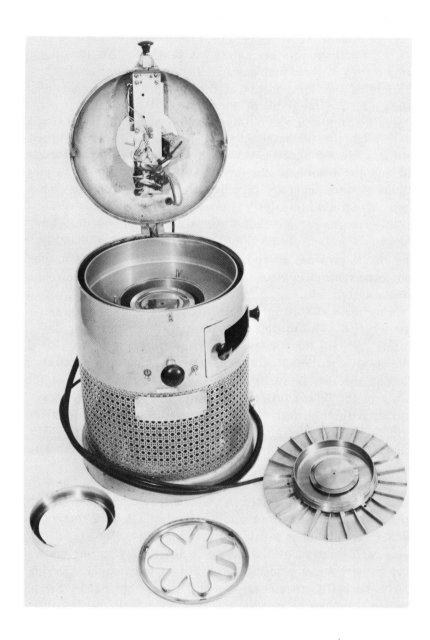

**Figure 43.** Disassembled view of microparticle classifier.

## TEST PROCEDURE

As a first step in the analysis of a new powder or material, examination under the optical microscope is highly recommended.    Even a brief examination can provide valuable information on the particle size range, particle shape and density, need for sieving, throttle spacer selection, and an awareness of any special problems or precautions such as friability, agglomeration and nesting that might be encountered and accommodated. Following the microscopic examination, the powder is oven dried at 105 to 110°C to constant weight and cooled and stored in a desiccator.    A sample is weighed to the nearest 0.01 g, and sieved through a 270* mesh sieve (53 $\mu$m) to provide 10 to 20 g of material less than 60 $\mu$m in size for the Bahco.    Record the weights of these samples as $W_S$ and $W_O$, respectively.

Place the less than 60 $\mu$m material in the Bahco feed hopper, insert and secure the largest throttle spacer (No. 18, corresponding to the smallest particle size fraction), and start the rotor motor.**    When the rotor reaches maximum speed, start the gas supply to the orifice, turn on the vibrator and brush motor, and adjust the vibrator to give the desired powder feed rate.    Check the feed rate at two 5-min intervals and adjust the vibrator as needed to maintain the desired flow rate. Once a feed rate of 1 g/min is established, readjustment will be needed only when the brush shows excessive wear.

Continue the operation until the hopper is empty, then switch off the vibrator, brush, orifice  gas supply  and the rotor motor.    Stop the rotor by gently applying the brake. When the rotor is completely stopped, swing the cover casting out of the way, remove the upper rotor assembly  and carefully place it on a large sheet of clean, glazed paper or aluminum foil.    Remove the catch basin and brush out the collected material, along with any adhering to the underside of the catch basin, onto the paper.    Weigh this material and record the weight as $W_{18}$.

---

*U.S. Sieve Series & Tyler Equivalents ASTM-E-11-61.

**Usual precautions for operating high-speed centrifuges must be exercised.

Reassemble the Bahco, replace the No. 18 throttle spacer with the next lower number and place the $W_{18}$ fraction of material in the feed hopper. Repeat the operation until all the throttle spacers have been used making sure to weigh the material collected in the catch basin with each spacer.

The material collected in the catch basin represents the coarse size fraction. The fine size fraction is deposited on the periphery of the rotor. It is not necessary to remove the fine fraction during the analysis, but it is often desirable, and necessary if the fractions are to be retained for some additional research or testing. A small fraction of the fine fraction should be examined under the microscope to determine the largest particle size present. The percent of material less than the size associated with the spacer is calculated as follows

$$F_{18} = 100(W_O - W_{18})/W_S \qquad (27)$$

Each powder should be submitted for density determination by an appropriate technique.

## CALIBRATION AND DETERMINATION OF
## TERMINAL SETTLING VELOCITY

Standard powder samples are available* for calibration of centrifugal classifiers. Instructions for use of the standards are enclosed. A graph showing the terminal settling velocity as a function of the size distribution is provided. A typical graph is shown in Figure 44. The powder is analyzed by the Bahco and the relationship between throttle spacer number and weight of the separated size fraction is established from Equation 27. The spacer numbers are plotted on the size distribution curve in Figure 44 and the terminal settling velocity values corresponding to each spacer are read from the ordinate of the graph.

---

*Write to Chairman, Powder Test Code Committee No. 28, c/o the American Society of Mechanical Engineers, 345 East 47th Street, New York, New York 10017.

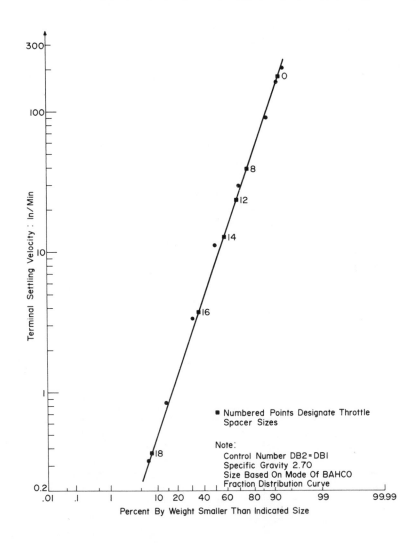

**Figure 44.** Terminal settling velocity distribution.

It is often convenient or required to report the size distributions as either equivalent unit density spheres or Stokes' diameters.* These transformations can be performed with aid of Figure 45. A typical report is given in Table XX.

---

*Stokes' diameter is the diameter of a sphere having the same density and terminal settling velocity as the particle.

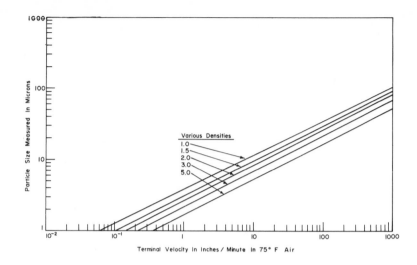

**Figure 45.** Terminal settling velocity vs spherical particle size as solid spheres.

**Table XX.** Bahco Size Analysis, Sample A

| Bahco Throttle Number | Terminal Settling Velocity (TSV) (in./min) | Cumulative Sample Weight Less Than or Equal to that TSV | Particle Size (µm) as Unit Density Spheres which Corresponds to that TSV |
|---|---|---|---|
| 18 | 0.37 | $F_{18} = 3.35\%$ | 2.2 |
| 16 | 4.6 | $F_{16} = 18.8\%$ | 8.0 |
| 14 | 26.0 | $F_{14} = 25.3\%$ | 19.0 |
| 12 | 41.5 | $F_{12} = 29.4\%$ | 25.0 |
| 8 | 95.0 | $F_{8} = 34.0\%$ | 36.0 |
| 4 | 152.0 | $F_{4} = 36.6\%$ | 45.0 |

# 11

# PROBLEMS IN PARTICLE SIZING:
# THE EFFECT OF PARTICLE SHAPE

Colin F. Harwood
  Pall Corporation
  30 Sea Cliff Avenue
  Glencove, New York 11542

## INTRODUCTION

There are many methods to measure the size of particulate solids. The many methods have been developed to give data on particles in relation to their state of subdivision, their origin, their chemical nature, their method of presentation to the sensor and to the intended use of the data obtained. In short, a method of measuring particle size which is ideally suited for one purpose may be totally inadequate for another application.

Here, the methods of measuring particle size will be considered in relation to the effect that the shape of the particle has on the measured result. In many instances, the irregular shape of a particle is an annoyance which leads to an interference in the accuracy of the results obtained. In other cases, the ability of an instrument to differentiate differences in shape is essential and is the reason for the instrument's development and use.

## MEASUREMENT OF PARTICLE SHAPE

Simple descriptive terms such as spherical, cylindrical, acicular, irregular and angular are easy to state. They can be readily applied to a particle, and the observer is rarely left in doubt as to the general form of the particle. Conversely, the actual scientific measurement of this simple descriptive term is extremely difficult in all but the most simple instances of perfectly symmetrical spheres, cylinders, cubes or blocks.

Once a particle assumes an irregular form, complex equations are required to define the shape relationship. Much pioneering work in this area has been performed by Heywood.[5][6] After studying thousands of grains of sand, he was able to derive the equation

$$f = 1.57 + C \frac{k}{m} 4/3 \frac{(n+1)}{n} \qquad (28)$$

where  $f$ = surface coefficient $(fd_a^2 = \pi d_s^2)$

$k$ = volume coefficient $(kd_a^3 = \pi d_v^3)$

$n = \cdot \dfrac{\text{length}}{\text{breadth}}$ = elongation ratio

$m = \dfrac{\text{breadth}}{\text{thickness}}$ = flatness ratio

$C$ = coefficient depending on geometric form.

The values of k and C for certain recognized geometric forms and also for some irregular particles are given in the following table:

Table XXI.  Shape Coefficients for Equidimensional Particles.

| Shape Group | | k | C | $Ck^{4/3}$ |
|---|---|---|---|---|
| **Geometrical forms:** | | | | |
| Tetrahedral | | 0.328 | 4.36 | 0.986 |
| Cubical | | 0.696 | 2.55 | 1.571 |
| Spherical | | 0.524 | 1.86 | 0.785 |
| **Approximate forms:** | | | | |
| Angular: | Tetrahedral | 0.38 | 3.3 | 0.91 |
|  | Prismodial · | 0.47 | 3.0 | 1.10 |
| Subangular | | 0.51 | 2.6 | 1.06 |
| Rounded | | 0.54 | 2.1 | 0.92 |

Another term used to help in the description of shape is sphericity.  As defined by Wadell,[57] the term relates $d_v$, the diameter of a sphere of equivalent volume, to $d_s$, the diameter of a sphere of equivalent surface.  Thus,

$$\text{Sphericity } \psi = \pi d_v^2 / \pi d_s^2$$

$$= \left( \frac{d_v}{d_s} \right)^2$$

The range of values of f, k  and $\psi$ for typical particles are given in Table XXII.

Table XXII.  **Range of Values of f, k  and $\psi$ for Various Materials.**

| Type of Material | k | f | $\psi$ |
|---|---|---|---|
| Rounded particles:   water worn sands, fused flue-dust, atomized metals | 0.32 to 0.41 | 2.7 to 3.4 | 0.817 |
| Angular particles of pulverised minerals, *e.g.*,  coal, limestone, sand | 0.2 to 2.28 | 2.5 to 3.2 | 0.655 |
| Flaky particles:   plumbago, talc, gypsum | 0.12 to 0.16 | 2.0 to 2.8 | 0.543 |
| Very thin flakes:   mica, graphite, aluminum | 0.01 to 0.03 | 1.6 to 1.7 | 0.216 |

## THE SIZE OF A PARTICLE

Because of the difficulties of describing the shape of a particle, as seen above, it is more usual to define the size of a particle by a single dimension, a diameter.  This can be accomplished by expressing the size of the particle in terms of the diameter of a sphere that is equivalent to the particle with respect to some stated property.  Any of the following equivalent spheres may be used for this purpose.

The diameter of the sphere which:

1. has the same projected area as the particle when viewed in a direction perpendicular to the plane of greatest stability, symbol $d_a$;

2. has the same volume as the particle, symbol $d_v$;

3. has the same surface area as the particle, symbol $d_s$;

4. has the same free-falling velocity in a fluid as the particle, symbol $d_f$ (if the diameter/velocity relationship follows Stokes' Law, the symbol $d_{st}$ is used); and

5. corresponds to a square aperture of side A through which the particle will just pass.

The variation between the above equivalent diameters increases as the particles diverge more from the spherical shape, and hence shape is an important factor in the correlation of sizing analyses made by differing procedures.

The above method of quoting a single number to determine particle size is convenient and, in many instances, sufficient. However, in many other instances it is insufficient, and notable examples would be fibers where an aspect ratio, length to breadth, is required; catalysts where an open, porous structure is often desirable and thus surface area is required; and paint pigments where flat particles with high hiding power are desirable.

## ANALYTICAL INSTRUMENTS IN RELATION TO THE PARTICLE SHAPE

There are a very large number of instruments which measure the size of particles. In Table XXIII the instruments have been arranged according to the fundamental parameter which they measure. These instruments will now be briefly reviewed with respect to the effect of particle shape on the results obtained.

Table XXIII.   Fundamental Particle Parameter Measured by
Several Instrumental Methods

| Parameter Measured | Analytical Method | Particle Size Range (μm) |
|---|---|---|
| Length | Screens | Down to about 40 |
| | Electroformed Screens | 50 to 10 |
| | Optical Microscope | 100 to 0.5 |
| | Electron Microscope | 5 to 0.001 |
| | Holography | 500 to 5 |
| Mass | Elutriation | 100 to 5 |
| | Sedimentation in Air | 200 to 5 |
| | Sedimentation in Liquid | 150 to 3 |
| | Centrifugal Sedimentation | 100 to 0.1 |
| Presented Area | Light Scatter | 50 to 0.3 |
| | Light Obscuration | 50 to 0.3 |
| Surface Area | Surface Area | $0.001 \text{ m}^2/\text{g}$ and up |
| | Permeametry | Mean Area Diameters from 0.01 to 100 |
| | Mobility Analyzer | 1 to 0.005 |
| | Condensation Nuclei Counters | 1 to 0.002 (total number only) |
| Volume | Electronic Sensing Zone | 200 to 0.2 |
| | Acoustic Particle Counter | 200 to 50 |

## Instruments Which Measure Length

### Microscope—Nonautomated

Images of the particles to be sized are focussed on an eyepiece graticule bearing a scale, circles, rectangles, etc.  A calibrated eyepiece micrometer may also be used, while larger
particles can be sized using the microscope stage.  Measurement by image-shearing is a more recent development.  The
operation of the microscope is skilled, tedious  and time con-
suming for reliable size analyses.  Qualitative microscopic
examination is, of course, less exacting and could well be
regarded as a necessary preliminary to any particle size analysis.

Operator fatigue imposes the principal limitation upon regular
use of the nonautomated microscope for size analysis.  At
least 300 to 500 particles require to be measured for an

approach to a meaningful analysis. On the other hand, for mixtures which cannot be separated into their components, visual microscope sizing is the only valid method in the subsieve size, or in any size range if the components have to be separately assessed. Here specialized techniques are occasionally useful such as the use of polarized light for identifying specific particles. Fluorescence microscopy and dark field illumination also find occasional application.

## Microscope—Automated

Automatic methods of counting and sizing have been developed and considerable progress is still being made in the application of electronics to microscope sizing. The complexities are reflected in the high prices of the instruments, which nevertheless become economic in terms of operator saving and often accuracy since more particles can be examined than would otherwise be practicable. Various principles of operation are employed. In one type of instrument an image of the particles is projected onto a slit with a conventional type of microscope. The sample is mechanically scanned past the slit and signals are electronically recorded.

Another type makes use of the image-splitting principle, while a more recent development utilizes the scanning spot of a television camera. This instrument gives a virtually instantaneous reading, and the size distribution can be derived by successive scans at different discrimination levels.

## Sieve Sizing

Hand sieving is both a highly skilled and tedious procedure. With the present availability of improved and mechanical sieving devices, it should be possible to standardize mechanical methods, possibly on a time-of-sieving basis, provided that the appropriate period has been previously determined by an end-point test. Particular attention should be paid to the condition of test sieves with due regard to distortion or wear of the mesh and also to the danger of attrition of the sample under test. Sieves should be examined before use and after use to ensure the integrity of the mesh. The accuracy of a sieving operation depends on factors such as the loading on the sieve, the application of the sieving movement and the shape of the particles.  In general, a sieve will allow particles to pass through based on their cross-section diameter rather than their length.

*Holography*

Modern developments in laser technology have made possible the use of laser holography.  Airborne particles may be photographed and their diffraction patterns recorded.  A three-dimensional image may then be reconstructed and information on the shape and size of the particles obtained.

*Electron Microscope*

Electron microscopes allow the observation of extremely small particles.  Transmission electron microscopes have a resolution limit of about 2 nm while scanning electron microscopes can resolve particles down to 200 to 500 nm.  The advantage of scanning electron microscopes is that they are able to give information on the topography of the particle as opposed to just a two-dimensional image.  Indeed, scanning microscopes usually have the ability to take stereo pairs from which considerable information on the depth of particles can be easily obtained.

## Instruments Which Measure Mass Differences

The mass of a particle will determine its movement in a fluid in which it is suspended.  Depending on whether the particle competes against an upward rising fluid stream or is falling through the still fluid by gravity, the principal is referred to as elutriation or sedimentation respectively.  The nomenclature has some confusion because versions of both elutriation and sedimentation instruments utilize centrifugal forces to increase the momentum of small particles.  This wide variety of instruments may utilize air or liquid as the suspending media, the separating forces may rely purely on gravitational effects or they may also involve centrifugal forces, the movement of the particles may be vertical, up or down, or they may be horizontal.

One problem common to all elutriators or sedimentation instruments is the need to adequately disperse the particles. Obviously particles which clump or agglomerate will have a higher effective mass than the individual particles. Prior to any experimental evaluation of particle size, studies should be made on the method of dispersing the particles, and this should be checked using a microscope.

The basic law governing the behavior of particles in a fluid stream is that of Stokes

$$D^2 = \frac{18 \, \eta v}{(\rho_s - \rho_f)g}$$

where   D = diameter of sphere

$\eta$ = viscosity of fluid

v = terminal velocity

$\rho_s$ = density of the sphere

$\rho_f$ = density of the fluid

g = acceleration due to gravity.

The following assumptions are made in the derivation:

1. The particles are smooth, spherical and rigid, and there is no slip between them and the liquid.

2. The particle must move as it would in a fluid of infinite extent.

3. The terminal velocity must have been reached.

4. The fluid must be homogeneous compared with the size of the particle.

5. The settling speed must be low so that only viscous forces are brought into play.

To compensate for the shape of the particles, it is necessary to use the factor k, the volume coefficient, as discussed previously.

Other common sources of error with this class of an instrument are particle interaction and circulation current. Particle

interactions are impossible to eliminate at the concentration
levels required for analysis.  The phenomenon of larger particles striking smaller particles as they fall at a faster rate is
known as hindered settling.  Circulation currents are initiated
by the movement of the particles as they pass through the
fluid and by convection currents due to cooling drafts or local
heating.

*Density Changes in Liquid Sedimentation (normally for*
*subsieve size)*

When a uniform dilute suspension settles, the composition at
a given depth is generally believed to remain unchanged until
some particles originally in the uppermost layer arrive.  These
will be the largest of that layer.  When they pass the given
depth, the concentration begins to fall and continues to decrease
as the suspension there becomes successively denuded of particles
of lesser size in turn.  Measurement of the rate of concentration
change allows the size distribution to be determined.

The concentration changes can be followed either gravimetri-
cally or optically.  Methods of following specific gravity changes
by hydrometer or divers (small sinkers of adjusted effective density) can only be regarded as approximative as they necessarily
obstruct the flow path of the particles.  Photometric methods
are established and convenient, and they have the advantage
that very low concentrations can be employed with corresponding reduction in interparticle interference.  The certainty
is limited by the theoretical assumptions, but for routine examination on a comparative basis, this objection does not hold.

*Sedimentation in Liquids (Centrifugal)*

Centrifugal sedimentation methods are still in a relatively
early stage of development, although a variety of commercial
instruments is available.  They extend the lower size range of
liquid sedimentation analysis down to 0.1 $\mu$m.  A recent
development is the disc centrifuge type of apparatus, in which
a rotating hollow disc is partly filled either with a homogeneous suspension or with clear suspending liquid on which a layer
of suspension is initially formed.  Samples are extracted for

gravimetric analyses, or the concentration changes are followed photometrically.

## Sedimentation in Air

Settling in air is more rapid than sedimentation in liquids under gravity.  In the Micromerograph, a sample of powder is blown through an annular slit between two mating conical surfaces at the top of a sedimentation column in order to disperse agglomerates.  The weight of powder settling on a pan is automatically recorded as a function of time, but some of the finer particles may be found to adhere to the long walls of the column.

## Elutriation

Elutriation is performed to classify small particles into size groups in the same manner as sieves are used for large sized particles.  The Bahco unit is probably best known because of its acceptance as the standard method for measuring pollutant dusts.  It utilizes centrifugal air classification.

## Instruments Which Measure Presented Area

Two types of instruments measure the interaction between a light beam and a particle.  One is relatively simple and measures the obscuration of the light beam by the particle placed in its path.  The second is complex and involves a measure of the scattered radiation after the light has struck the particle.  The scattered radiation contains components which are diffracted, refracted and reflected from the particle and also includes radiation which has been adsorbed and retransmitted at a larger wavelength.

The theory of light scattering is very complex.  Boundary conditions on the theory limit the application to spherical particles in the size range 0.1 to 4.0 $\mu$m.  Extensions to this theory can allow particles up to 50 $\mu$m to be included to some degree.  Thus, it can be seen that the utility of these devices is strictly limited for particles of irregular shape and wide size range.

Despite these limitations, the devices do find considerable utility. Their chief advantage is in giving an instantaneous reading of the number of small particles in a given volume of fluid. For this reason, they are commonly used to check the particle content of air or liquid that has been ultrafiltered to a very low contamination state.

### Instruments Which Measure Surface Area

The specific surface of a powder or bulk solid material is the surface area of the component particles in unit weight. The surface-mean-diameter of a powder is defined as the size of a particle of the same material which has the same specific surface as the powder, but the true value can only be derived from the latter if the appropriate shape factor is known. The values usually given are generally based on the assumption of spherical particles.

For some purposes, the effective size characteristic of a particulate is its specific surface, and a knowledge of size distribution is not essential. Examples are found in catalysis and the evaluation of crushing energy. There are two quite different approaches to the determination of specific surface, one based on permeability or resistance to flow of a fluid, when the material is compacted, while the other is based on gas adsorption. The approaches could not be expected to give the same results: the one measures in effect the "envelope" area of the particles, while the other yields the "absolute" area.

With permeability methods, the size characteristics are inferred from the resistance offered to the flow of a fluid through a pressed plug of powder. The Kozeny-Carman equation is used to estimate the average particle size and surface area. The results obtained rarely agree with those obtained from instruments which measure surface area by adsorption. The instrument is, however, easy to operate and inexpensive.

True surface area is measured by gas adsorption. The quantity of gas required to form a monolayer on the powder surface is measured. Since the area of cross section of the gas molecules is known, the surface area of the powder is then calculated. There are a number of equations relating to gaseous adsorption. The theory generally used is that of Brunauer, Emmet, and Teller, and hence the method is referred to as the BET method.

Other devices utilizing the surface properties of powder are the mobility analyzer and the condensation nuclei counter. Neither of these devices are sensitive to the shape of the particle. Both measure aerosol particles which are very low in particle size and hence have utility in establishing the efficiency of ultimate filters.

The mobility analyzer operates on the principle that a given sized particle will achieve a given mobility in an electrical field when charge under standard conditions. A series of tests of a given aerosol will quickly obtain a size distribution based on the surface area available for charging.

The condensation nuclei counter is a simpler, less expensive device which will give information on the number of small particles present. Here, the principle is that particles will act as nucleation sites when placed in a high-humidity chamber. A measure of the light obscuration after nucleation leads to information on the number of particles present in the sample.

## Instruments Which Measure Volume

The volume of a particle may be measured by passing it through a sensing zone and observing the size of a pulse generated. It is essential that the number of particles is small so that the particles pass through the zone one at a time, otherwise a signal which is larger than it should be will be recorded. This type of error is known as coincidence error.

The best known instrument of this class utilizes an electronic sensing of particles held in an electrolyte which pass through an orifice creating a voltage pulse. To a large extent, the results are independent of particle shape provided the shapes are not extreme. The instrument can be deliberately adapted to measure particle shape of, for example, fibers by measuring the length of the pulse in addition to the height of the pulse.[58]

Acoustic particle counters have also been developed. These instruments work on a complex and, at present, little understood theory. In essence, airborne particles in the size range of 200 to 50 μm are fed into a specially shaped tube.  The particles are accelerated and produce an audible click which can be picked up by a microphone, amplified and recorded.  They can give no indication of the size or shape of the particles within their operating range.

## CONCLUSIONS

It is shown that the measurement of particle shape is extremely difficult for any particle other than the most simple forms.  If it is necessary to measure the particle shape, an instrument which will allow the direct measurement of length, breadth, depth and angularity is required.  Such instruments include optical microscopes and electron microscopes, the readings from which may be automated to some degree by use of an image-analyzing computer.  Holography offers a relatively new technique for assessing the size and shape of airborne particles.

All other techniques of measuring particle size are affected by the shape of the particle but do not lead to information of the shape itself.  Thus, instruments which measure the particle size by a technique which utilizes the effect of particle mass will be in error due to the divergence from Stokes' Law.  This divergence will, in terms of particle shape, be caused by lack of sphericity and by the pore structure.  Methods which utilize a measurement on the presented area are inherently limited in terms of irregular particles.  The particle size estimated from surface area measurement will be in large error if the particle has a very open pore structure, giving rise to a large internal surface.  Instruments which measure particle volume will not generally be affected by shape unless the shape is extreme.  In special circumstances, the instrument can be adapted to measure the shape of particles by monitoring pulse length and height.

12

# SELECTION OF A PARTICLE
# SIZE MEASUREMENT INSTRUMENT

Edward G. Fochtman
IIT Research Institute
10 West 35th Street
Chicago, Illinois 60616

## INTRODUCTION

The previous chapters of this book have outlined techniques of presenting particle size data and the basic principles of many instruments used to measure particle size. Once the type of particle size information required for a given application has been established, it is necessary to select the instrument most suitable for the development of this information.

The selection of the best particle size measuring instrument for a given application is a difficult task and careful review of the requirement and instrument capabilities is advised. In many cases, trial use of an instrument or of data from a firm providing a particle size measurement service would be helpful.

As an aid to selecting an instrument, the following general guidelines may prove useful.

There are three basic steps in the selection of an instrument for particle size analysis:

    1.    Definition of requirements

    2.    List the equipment alternatives

3.   Select the equipment on the basis of initial cost, time for analysis, degree of automation, and sample size.

There are two general considerations in the measurement of particles that are not strictly instrument-oriented but are essential to the development of meaningful data. First, it is necessary to obtain a representative sample. In general, the instrument will examine only an extremely small portion of the total and the sample should be withdrawn to be as representative as possible. Second, the manipulations on the sample should be minimized since every operation introduces the opportunity for change in the sample with the attendant erroneous results.

## DEFINITION OF REQUIREMENTS

The first step in the selection of an instrument is the development of a good definition of the requirements.

The system to be analyzed must be characterized relative to the dispersion, the particle size range, the particle shape, particle density and physical and chemical nature of the material.

The dispersion may involve either a liquid or solid in a gas, or a liquid or solid in a liquid. Since it is best to minimize manipulation of the sample, an instrument which will measure the dispersion directly should be sought. If this is not possible, the method of collecting the sample and presenting it to the instrument must be such that its integrity is maintained.

The particle size range to be examined often dictates the type of instrument which can be used. For convenience, the size ranges can be considered as:

- less than 1 $\mu$m
- 1 to 50 $\mu$m
- greater than 50 $\mu$m

Definition of particle shape may be an essential aspect of the required data and is important as the selection of an instrument which either develops this information or is compatible with the requirements. Shape factors often become important in the identification of particulate materials and in the study of abrasion or wear.

In several types of instruments, the particle density is used in the determination of particle size. Although the density can be determined from handbooks for solid nonporous particles, it is much more complex when one considers the flocculant particle, the porous particle such as a catalyst particle, or the fly ash type particle which has voids caused by gas bubbles. Obviously measuring these types of particles with an instrument based on particle density will produce information which must be used with considerable caution.

Many other physical and chemical characteristics of the particles should be considered. Properties such as transparency or refractive index are used by some instruments. Melting point and vapor pressure become important in electron microscopy. The frangibility of particles may be important in the preparation of samples for examination. Particles that are gellike present problems for some types of separating and measuring instruments.

## EQUIPMENT ALTERNATIVES

The microscope is a basic tool for all laboratories concerned with fine particles. Even a simple microscope which is available at nominal cost can provide a wealth of useful information.

Once the requirements of the analysis for a particular application have been defined, it is possible to select one or more instruments which will fulfill the requirements. Two different types of instruments may be needed.

In addition to the purely technical aspects of the requirements, there may be need for permanent records, rapid reporting of results, or minimization of the dependence upon operator technique. The basic instrument used for sensing particle size can often be equipped with accessories which will make permanent records and/or perform certain calculations and present results in tabular or graphic form.

All of the alternatives should be investigated and tabulated for comparison purposes.

Table XXIV. Summary of Particle Size Instrumentation Characteristics

| Initial Cost, $ | Type | Size Range (μm) | Sample Capability | | | | | Special | Approximate Analysis Cost ($) |
|---|---|---|---|---|---|---|---|---|---|
| | | | Liq/Gas | Solid/Liq | Liq/Liq | Solid/Gas | Solid | | |
| 500-1,000 | Optical Microscope | 1 to 1,000 | No | Yes | Yes | Yes | Yes | Shape | 40 |
| | Sieves | 1 to 1,000 | No | Yes | No | Yes | Yes | | 20 |
| | Simple Sedimentation | 5 to 1,000 | No | Yes | Yes | Yes | Yes | Density | 40 |
| | Gas Permeability | 5 to 1,000 | No | No | No | Yes | Yes | Surface | 10 |
| 10,000 | Optical Microscope | 0.5 to 1,000 | No | Yes | Yes | Yes | Yes | Shape | 40 |
| | Electron Microscope | 0.2 to 10 | No | Yes | No | Yes | Yes | Shape | 20 |
| | Centrifugal Sedimentation | 0.2 to 50 | No | Yes | Yes | Yes | Yes | Density | 20 |
| | Sedimentation | 0.2 to 50 | No | Yes | Yes | Yes | Yes | Density | 20 |
| | Gas Adsorption | 0.2 + | No | No | No | Yes | Yes | Surface | 40 |
| 10,000-100,000 | Electron Microscope | 0.1 to 10 | No | Yes | No | Yes | Yes | Shape | 40 |
| | Scanning Electron Microscope | 0.1 to 100 | No | Yes | No | Yes | Yes | Shape, 3D | 200 |
| | Stream Counting | 1.0 to 100 | Yes | Yes | Yes | Yes | Yes | Rapid | 10 |
| | Instrumented Microscope | 0.5 to 1,000 | No | Yes | Yes | Yes | Yes | Rapid | 20 |
| 100,000 + | Scanning Electron Microscope | 0.1 to 100 | No | Yes | No | Yes | Yes | Shape, 3D | 200 |

## ECONOMIC CONSIDERATIONS

The economic evaluation of an instrument must consider the value of the data as well as the cost of obtaining the information. Modern instrumentation often provides information which increases the ability to control processes and increases uniformity of product and decreases rejected material. In addition, instrumentation often frees an operator from a tedious job, and produces information on a rapid and consistent basis.

Cost of particle size instruments increases rapidly as the particle size to be examined decreases. Cost of instruments range from about $500 to well over $100,000.

A summary of instrument capabilities and capital cost range, along with the cost for semiroutine sample analysis, is given in Table XXIV. This serves as an initial guide in the selection of instruments for a particle size analysis laboratory; however, in-depth studies should be undertaken before committing large expenditures for instrumentation.

# REFERENCES

## CITED REFERENCES

1. White, H. *Industrial Electrostatic Precipitation* (Reading, Mass: Addison-Wesley, 1963).
2. Allen, T. *Particle Size Measurement* (London: Chapman and Hall Ltd.), London EC4, Britain.
3. Hatch, T. and S. Choate. *J. Franklin Institute 207:*369 (1929).
4. *Air Pollution Manual,* Part II Control Equipment, (Detroit: 1968) American Industrial Hygiene Association.
5. Kaye, B. *Chem. Eng. 73:*239 (1966).
6. Stockham, J. D. and M. R. Jackson. *Humidification and Mist Therapy, Fundamental Concepts of Aerosols,* Vol. 8 (3) (New York: Little, Brown, and Little, 1970), pp. 685-726.
7. Mercer, T. T. *Aerosol Technology in Hazard Evaluation,* Chapter 3 (New York: Academic Press, 1973).
8. Delly, J. *The Particle Analyst,* Reticles and Graticules, Vol. 1, No. 11, (Ann Arbor, Mich: Ann Arbor Science Publishers, Inc., 1965).
9. Silverman, L., C. Billings, and M. First. *Particle Size Analysis in Industrial Hygiene* (New York, N.Y.: Academic Press, 1971).
10. Chamot, B. and Mason, C. *Handbook of Chemical Microscopy,* Vol. 1, 3rd Ed., (New York, N.Y.: John Wiley & Sons, 1958).
11. "The Industrial Use of the Microscope," (Chicago, Ill: McCrone Research Institute, July 1962).
12. Stockham, J. "What is Size: Relationship Among Statistical Diameters," *Filtration Day 1975,* Midwest Chapter of the Filtration Society, IIT Research Institute, Chicago, Ill. (October 24, 1975).
13. Knutson, E. "Methods of Data Presentation," *Filtration Day 1975,* Midwest Chapter of the Filtration Society, IIT Research Institute, Chicago, Ill. (October 24, 1975).
14. "Methods for the Determination of Particle Size of Powders," *Optical Microscope Method,* BS 3406: Part 4 (1963), British Standards Institution, 2 Park St., London, W.1. England.

131

15. American Society for Testing and Materials. "Recommended Practice for the Analyses by Microscopical Methods," ASTM: Part 23 and 30, Philadelphia, Pa., 1973.
16. Ashley, H. "Centrifugal Analyzer - Bahco," *Filtration Day 1975,* Midwest Chapter of the Filtration Society, IIT Research Institute, Chicago, Ill. (October 24, 1975).
17. Fisher, C. *Microscope 19:*1 (1971).
18. Q-720 Users School given by Imanco, Monsey, New York (1973).
19. Wadlow, E. E., B. M. Hopkins, G. M. Gardner and C. Fisher. *Microscope 20:*183 (1972).
20. Cole, M. *Microscope 19:*87 (1971).
21. Gibbard, D. W., D. J. Smith and A. Wells. *Microscope 20:*37 (1972).
22. *Image Analyzing Computers:* Quantimet 720 Instruction Manual, (Hertfordshire, England, 1971).
23. Kaye, B. H. "Sedimentation Balance Methods of Particle Size Analysis," *Paint, Oil, Colour J.* (July 1966).
24. Kaye, S. M., D. E. Middlebrooks and G. Weingarten. "Evaluation of the Sharples Micromerograph for Particle Size Distribution Analysis," Technical Report F.R.L.-T.R.-54, Picatinny Arsenal, Dover, N.J. (February 1962).
25. Bostock, W. "A Sedimentation Balance for Particle Size Analysis in the Subsieve Range," *J. Sci. Instruments 29:*209 (1952).
26. Van de Hulst, H. C. *Light Scattering by Small Particles* (New York: Wiley, 1957).
27. Kerker, M. *The Scattering of Light and Other Electromagnetic Radiations* (London: Academic Press, 1969).
28. Davies, R. *Amer. Lab.* (December, 1973) p. 73.
29. Karuhn, R., R. Davies, B. H. Kaye and M. J. Clinch. *Powder Technology 11:*157-171 (1975).
30. Schmitz, J. E. *Filt. Sep.* 426-428 (1973).
31. "Coulter Counter Model B Instruction Manual," Coulter Electronics Inc., Hialeah, Florida.
32. "Submicron Separation and Data," IITRI Report No. C6239-A008, (June 1974), IIT Research Institute, Chicago, Illinois 60616.
33. "LFE Model 302 Low Temperature Asher Instruction Manual," LFE Corporation, Waltham, Mass.
34. "Millipore Catalogue and Purchasing Guide," Millipore Corporation, Bedford, Mass.
35. "Specifications and Physical Properties," Nuclepore Corporation, Pleasanton, Calif.
36. Goodhue, L. D. and C. M. Smith. *Ind. Eng. Chem., Anal. Ed. 8:*469-472 (1936).
37. Bouyoucos, G. J. *Soil Science 23:*343-349 (1927).
38. Oden, S. *Soil Science 19:*1-35 (1925).

39. Romwater and Vandl. *Colloidzchr 72:*1 (1935).
40. Slater, C. and L. Cohen. *J. Sci. Instrum 39:*614-617 (1962).
41. Burt, M. W. G. and B. H. Kaye. *The Analyst 91*(1086):547-552 (1966).
42. May, K. R. *J. Sci. Instrum. 22:*187-195 (1945).
43. Wilcox, J. D. *A.M.A. Arch. Ind. Hyg. Occup. Med. 7:*376 (1953).
44. Davies, C. N. and M. Alyward. *Proc. Phys. Soc. (London) B 64:* 889 (1951).
45. Ranz, W. E. and J. B. Wong. *A.M.A. Arch. Ind. Hyg. and Occup. Med. 5:*464 (1952).
46. Mercer, T. T., M. I. Tillery and G. J. Newton. *Aerosol Science 1*(9) (1970).
47. Marple, V. A. and B. Y. Liu. *Environ. Sci and Tech. 8*(7):648-654 (July 1974).
48. Couchman, J. D. "Use of Cascade Impactors for Analyzing Airborne Particles of High Specific Gravity," *Conf.-650407, 2:*1162, TID-4500 (1965).
49. Laskin, S. in: *Pharmacology and Toxicology of Uranium Compounds I,* C. Voegtlin, and H. C. Hode, Eds. (New York: McGraw-Hill, 1949) p. 463.
50. Lippman, M. *Amer. Ind. Hyg. Ass. J. 20:*406 (1959).
51. Andersen, A. A. *J. Bacterid 76:*471-484 (1958).
52. Davies, C. N., M. Alyward and D. Leacy. *A.M.A. Arch. Ind. Health Occup. Med. 4:*354-397 (1951).
53. "Determining the Properties of Fine Particulate Matter, Powder Test Code #28," *American Society of Mechanical Engineers,* United Engineering Center (New York, N.Y.: ASME)
54. "Microparticles Classified Centrifugally," Bulletin 56-52, Harry W. Dietert Co., Detroit, Michigan
55. "Instructions for No. 6000 Bahco Microparticle Classifier," Harry W. Dietert Co., Detroit, Michigan
56. Heywood, H. *Symposium on Particle Size Analysis,* London, 1947.
57. Wadell, H. "Volume, Shape, and Roundness of Rock Particles," *J. Geol. 40:*443 (1932).
58. Davies, R., R. Karuhn, J. Graf and J. Stockham. "Measurement of Fiber Size Distribution in Parenteral Solutions," *Bul. Parenteral Drug. Assoc. 29*(2):110 (1975).

## UNCITED REFERENCES

Alex, W. "Principles and Classification of Counting Methods in Particle Size Analysis," *Aufbereits-Technik 13:*2,3,10,11 (1972).
Alex, W. *et al.* "Particle Size Analysis: 4, Enumeration Methods," *Chem-Ing-Tech 46*(11):477 (1974).

"American National Standard Method for Calibration of Liquid Automatic Particle Counters Using AC Fine Test Dust," American National Standards Institute, Std. B93.28-1973 (1973).

Anderson, F. G. *et al.* "Analyzing Midget Impinger Dust Samples with an Electronic Counter," Bureau of Mines Report of Investigation No. 7105 (April 1968).

Anduze, R. A. and J. C. Harris. "Methods for Particle Counting in Hydraulic Fluids," Monsanto Research Corp., Dayton, Ohio, Contract No. AF33 616 8438, Project 7381, Task 738103, September 1961-November 1963, May 1964.

Bader, H., A. R. Gordon and O. B. Brown. "Theory of Coincidence Counts and Simple Practical Methods of Coincidence Count Correction for Optical and Resistance Pulse Particle Counters," *Rev. Sci. Inst.* *43*(10):1407 (1972).

Bensh, L. E. and E. C. Fitch. "The Analysis of Particulate Contaminants in Hydraulic Fluids," Proceedings 28th NFPA Conference, Chicago (September 13, 1972).

Bowing, H. A. and T. Gast. "A New Method of Rapid Particle Size Analysis in the Range 0.1 to 10 μm," *Staub* (English), *30*(11): 15 (1970).

Burt, M. W. G. "Accuracy and Precision of the Centrifugal-Disc Photosedimentometer Method of PSD Analysis," *Powder Technol.* *1:*103 (1967).

Carr-Brion, K. G. and P. J. Mitchell. "An On-Stream X-Ray Particle Sensor," *J. Sci. Instr.* *44*(8):611(1967).

Carver, L. D. "Particle Size Analysis," *Ind. Research.* (August 1971) p. 40.

Connor, P. "Automatic Counting and Sizing of Particles," *Ind. Chemist* (February 1963) p. 69.

Corn, M. "Statistical Reliability of PSD Determined by Microscopic Techniques," *AIHA J 26:*8 (1965).

Curby, W. A. "Applications for the Multichannel Particle Size Analyzer," *Bull. Parenteral Drug. Assn.* *23*(5):208 (1969).

Davies, R. "Rapid Response Instrumentation for Particle Size Analysis, Part I," *Am. Lab.* (December 1973) p. 17.

Davies, R. "Rapid Response Instrumentation for Particle Size Analysis, Part II," *Am. Lab.* (January 1974) p. 73.

Davies, R. "Rapid Response Instrumentation for Particle Size Analysis, Part III," *Am. Lab.* (February 1974) p. 47.

Flinchbaugh, D. A. "Expended Capability of the Coulter Counter with a New Aperture Apparatus," *Anal. Chem.* *43*(2):172 (1971).

Jahr, J. "A New Sedimentation Procedure for Dust Samples," *Staub* (English) *30*(11):27 (1970).

John, P. T. and J. N. Bohra. "Particle Size Measurement of Industrial Powders," *J. Sci. Ind. Res.* *28*(3):85 (1969).

Joy, A. S., *et al.* "On-Stream Methods of Particle Size Analysis," *Proc. Soc. Anal. Chem 5*(5):80 (1968).

Karhnak, J. M. "A Realistic Roll-Off Cleanliness Procedure for Fluid Power Systems," Presented at the 29th National Conference on Fluid Power, Cleveland, Ohio (September 1973) pp. 25-27.

Kime, H. B. "Measuring the Retention of Filter Paper," *Filt. and Sep.* 2(4):296 (1965).

Kinsman, S. "Instrumentation for Filtration Tests," Proc. 78th AIChE Meeting, Salt Lake City, August 19, 1974; *Chem. Eng. Prog.* 70(12): 48 (December 1974); *Filt. Sep.* 12(4):376 (July/August 1975).

Kirnbauer, E. A. "Contamination Measurement in Hydraulic Fluids," Proc. 25th NCFP Conf., (Pall FSR No. 44), October 15, 1969.

Lapple, C. E. "Particle-Size Analysis and Analyzers," *Chem. Eng.* 75(11): 149 (1968).

Lieberman, A. "Problems in Handling and Analyzing Liquids with Low Levels of Particles," Proc. Eurocontamination Conference, Paris, October 1971.

Lien, T. R. and C. R. Phillips. "Determination of Particle Size Distribution of Oil-in-Water Emulsions by Electronic Counting," *Environ. Sci. Technol.* 8(6):558 (1974).

Lloyd, P. J. and A. S. Ward. "Filtration Applications of Particle Characterization," *Filt. Sep.* 12(3):246 (May/June 1975).

Martens, A. E. and R. Morton. "Automatic Image Analysis: A New Technique to Enhance an Old Art," *Canadian Res. Devel.* (January/February 1973).

Nakajima, Y. and T. Tanaka. "Design of the On-Line Particle Size Detector for Fine Powders," *I&EC Fund.,*10(2):318 (1971).

Olivier, J. P. *et al.* "Rapid Automatic Particle Size Analysis in the Subsieve Range," *Powder Technol.* 4(5):257 (1970/71).

Osborne, B. F. "A Complete System for On-Stream Particle Size Analysis," Proceedings CIM Conference Metallurgists, Montreal (1971).

"Particle Counting Technology," *AACC J.* 8(1):11 (1969).

Pietsch, W. B. "An Evaluation of Techniques for PSD Analysis, II," *Minerals Proc.* (December 1968) p. 12.

Scarlett, B. "Particle Size Analysis," *Filt. Sep.* 2(3):215 (1965).

Schmitz, J. E. "Particle Size Analysis of Oil Samples with a Coulter Counter," *Filt Sep.* 10(4):426 (1973).

Schrag, K. R. and M. Corn "Comparison of Particle Size Determined with the Coulter Counter and by Optical Microscopy," *AIHA J.* 31(4):446 (1970).

Sparrow, E. "Evaluation of Liquid-Borne Particle Monitor," Proceedings 4th AACC Meeting, Miami, May 25, 1965.